Scientific Explanation and
the Causal Structure of the World

Scientific Explanation and the Causal Structure of the World

Wesley C. Salmon

Princeton University Press
Princeton, New Jersey

Copyright © 1984 by Princeton University Press
Published by Princeton University Press, 41 William Street,
Princeton, New Jersey 08540
In the United Kingdom: Princeton University Press,
Guildford, Surrey

All Rights Reserved

Library of Congress Cataloging in Publication Data will be found on the
last printed page of this book

ISBN (cloth) 0-691-07293-0
(LPE) 0-691-10170-1

This book has been composed in Linotron Times Roman

Clothbound editions of Princeton University Press books are printed
on acid-free paper, and binding materials are chosen for strength
and durability. Paperbacks, although satisfactory for personal
collections, are not usually suitable for library rebinding

Printed in the United States of America
by Princeton University Press, Princeton, New Jersey

To
Carl G. Hempel
Philosophical Master

aka
Peter Hempel
Colleague and Friend

Contents

Preface

ALTHOUGH there is no doubt in my mind that scientific understanding of the world and what transpires within it represents a marvelous intellectual achievement, the chief purpose of this book is not to laud scientific understanding, but, rather, to try to get clear on what it is. Our aim is to understand scientific understanding. We secure scientific understanding by providing scientific explanations; thus our main concern will be with the nature of scientific explanation.

The attempt to find an adequate characterization of scientific explanation is a problem that has perplexed scientists and philosophers for several millennia. As we shall see in chapter 1, its roots go back at least to Greek antiquity, and it continues to be an object of considerable attention today. There are, it seems to me, three basic conceptions that have figured significantly in the discussions for twenty-five hundred years. We shall examine them in some detail.

One particular view has played a dominant role, at least since the middle of the twentieth century, in the philosophical literature on scientific explanation. It has often been called—somewhat inappropriately I think—the ''covering law model.'' Its crucial tenet is that explanations are arguments. Another view, which has enjoyed some philosophical support, has a good deal of common-sense appeal. It takes scientific explanations to be essentially causal. A major aim of our investigations will be to draw a sharp contrast between these two conceptions. A third conception, which inseparably associates explanation with necessitation, will be judged scientifically anachronistic.

It will be my contention that the 'received view' (the inferential conception) is infected with severe difficulties that, in the end, render it untenable. The causal conception also suffers from serious problems—namely, philosophical perplexities concerning the nature of causality—but I believe they can be overcome. This can be done only if we are able to provide an adequate philosophical treatment of fundamental causal concepts, as I attempt to do in chapters 5–7.

The distinction between these two conceptions of scientific explanation is by no means trivial or merely verbal. As I indicate in the final section of the last chapter, the transition from the one conception to the other demands nothing less than a radical gestalt switch in our thought about scientific explanation. I shall, of course, attempt to provide strong philosophical motivation for making the shift.

Although much modern work on scientific explanation has been rather formal and technical—often treating various quasi-formal 'models' in great detail—I shall dwell extensively upon less formal considerations. There are two reasons for this emphasis. In the first place, I have been convinced for some time that many recent philosophical discussions of scientific explanation suffer from a lack of what Rudolf Carnap called "clarification of the explicandum." As Carnap has vividly shown, precise philosophical explications of important concepts can egregiously miss the mark if we do not have a sound prior informal grasp of the concept we are endeavoring to explicate. It is impossible, I think, to read chapters 1, 2, and 4 of Carnap's classic, *Logical Foundations of Probability* (1950), without seeing the profound importance of such preliminary clarifications. Because of my strong conviction that the concept of scientific explanation needs similar elucidation, chapters 1, 4, and 9 of this book are largely devoted to this task.

In the second place, I have earnestly attempted to make the discussion accessible to a wide group of readers—philosophers, scientists, and other individuals who have serious interests in science or philosophy. Every effort has been made to illustrate the philosophical points with examples that have either historical or contemporary significance from a wide range of scientific disciplines. At the same time, a number of examples have deliberately been drawn from such sources as *Scientific American* and *Science*, for I hope this book will be intelligible to those who find such literature rewarding. No particular specialized knowledge of any scientific discipline is presupposed. In addition, I have tried to supply enough philosophical background to obviate any need for prior familiarity with the philosophical literature on scientific explanation. It is my special hope that some parts of this book—for example, the discussions of causality—will be useful to philosophers who work in areas outside of philosophy of science.

I do not mean to suggest that this book is easy; it is not. For the most part, however, it should be comprehensible to intelligent readers, more or less regardless of background, who are willing to expend some effort upon it. Unavoidably, some parts are rather technical—in particular, the details of the S-R basis in chapter 2 and the treatment of randomness and objective homogeneity in chapter 3. These can be skimmed or omitted on first reading by those who are not interested in such technicalities for their own sake.

This book had its real beginnings when, around 1973, I began to recognize the serious shortcomings and limitations of the statistical-relevance model of scientific explanation—as propounded in (Salmon et al., 1971)—and the need to come to grips with the explanatory force of scientific theories. At his kind invitation, I promised to contribute a paper entitled "Theoretical Explanation" to a symposium on explanation organized by Stephan Körner at Bristol; the proceedings are published in (Körner, 1975). As sometimes happens when the title is written before the paper, the content of the paper did not fit the title very well; all of my efforts to deal with theories got bogged down in problems of causality. As chapters 5–8 of the present book will show, I now believe that causal explanation is the key to theoretical explanation. No adequate account of theoretical explanation could be developed without extensive work on causality itself.

The actual writing of the book began during a visit to the Department of History and Philosophy of Science at the University of Melbourne, where I was invited to give a seminar on scientific explanation in 1978. Suffice it to say that every chapter bears the marks, explicitly or implicitly, of stimulating interactions with Australian philosophers. I am indeed grateful for the personal and intellectual hospitality accorded me during a splendid three-month visit.

There are several individuals to whom I owe special thanks for substantive help in developing the ideas in this book. Nancy Cartwright pointed out a serious flaw in an earlier attempt to explicate the propagation of causal influence, and stimulated me to improve my account. Alberto Coffa made me see the crucial role of epistemic relativization in Carl G. Hempel's theory of inductive explanation, and made me realize the indispensable character of objective homogeneity for the type of theory I was attempting to develop. Clark Glymour called to my attention the epoch-making work of Jean Perrin on Avogadro's number and directed me to Mary Jo Nye's superb account of that work (1972). Paul Humphreys offered substantial aid in dealing with objective homogeneity, probabilistic causality, and propensities. Patrick Maher and Richard Otte noticed a fatal flaw in an earlier explication of causal interaction and forced me to correct that account. Otte also provided valuable critiques of various aspects of probabilistic causality. Hugh Mellor, more than anyone else, led me to appreciate the importance of clarifying the explicandum in discussions of scientific explanation. Bas van Fraassen made me realize that there must be two kinds of causal forks, conjunctive and interactive. Philip von Bretzel first called to my attention the direct relationship between causal forks and causal interactions.

I cannot begin to detail the ways in which my ideas on scientific explanation have been influenced by the writings of Bertrand Russell—especially *The Analysis of Matter* (1927), *Mysticism and Logic* (1929), and *Human Knowledge, Its Scope and Limits* (1948)—as well as Hans Rei-

Acknowledgments

I AM GRATEFUL to *Scientific American* for permission to use the excerpts from "Quantum Theory and Reality" by Bernard d'Espagnat, Copyright © 1979 by Scientific American, Inc. These excerpts appear in chapters 1 and 9; specific page references are given at the end of each quotation.

I should like to express my sincere appreciation to the following publishers for permission to adapt material from my previously published articles for use in this book:

Portions of the material in chapters 1 and 4 were included, in a much abbreviated form, in "Comets, Pollen, and Dreams: Some Reflections on Scientific Explanation," in Robert McLaughlin, ed., *What? Where? When? Why?*, pp. 155–178. Copyright © 1982 by D. Reidel Publishing Company, Dordrecht, Holland.

The section "Epistemic Relativization" in chapter 3 is adapted from "Indeterminism and Epistemic Relativization," *Philosophy of Science* 44, pp. 199–202. Copyright © 1977 by the Philosophy of Science Association, East Lansing, Michigan.

The sections "Randomness" and "Homogeneity" in chapter 3 are drastically revised versions of material that appeared in "Objectively Homogeneous Reference Classes," *Synthese* 36, pp. 399–414. Copyright © 1977 by D. Reidel Publishing Company, Dordrecht, Holland.

The section "The Epistemic Conception" in chapter 4 includes material adapted from "A Third Dogma of Empiricism," in Robert Butts and Jaakko Hintikka, eds., *Basic Problems in Methodology and Linguistics*, pp. 149–166. Copyright © 1977 by D. Reidel Publishing Company, Dordrecht. It also contains material from "Hempel's Conception of Inductive Inference in Inductive-Statistical Explanation," *Philosophy of Science* 44, pp. 180–185. Copyright © 1977 by the Philosophy of Science Association, East Lansing, Michigan.

Some of the basic ideas in chapters 5 and 6 were presented in a preliminary and much-condensed version in "Causality: Production and Propagation," in Peter Asquith and Ronald N. Giere, eds., *PSA 1980*, pp. 49–69. Copyright © 1982 by the Philosophy of Science Association, East Lansing, Michigan. The section "The 'At-At' Theory of Causal Propagation" in chapter 5 is an adaptation of "An 'At-At' Theory of Causal

Influence," *Philosophy of Science* 44, pp. 215–219. Copyright © 1977 by the Philosophy of Science Association, East Lansing, Michigan.

Most of the material in chapter 7 was drawn from "Probabilistic Causality," *Pacific Philosophical Quarterly* 61, pp. 50–74. Copyright © 1980 by the University of Southern California.

Some of the material in chapters 8 and 9 was anticipated in a much less complete form in "Further Reflections," in Robert McLaughlin, ed., *What? Where? When? Why?*, pp. 231–280 (especially pp. 260–278). Copyright © 1982 by D. Reidel Publishing Company, Dordrecht, Holland.

I should like to express special thanks to the University of Melbourne for providing me with a research appointment in September through November 1978, and to the National Science Foundation for support of my research on causality and scientific explanation.

This material is based upon work supported by the National Science Foundation under Grant No. GS-42056 and Grant No. SOC-7809146.

Any opinions, findings, and conclusions or recommendations expressed in this publication are those of the author and do not necessarily reflect the views of the National Science Foundation.

Scientific Explanation and
the Causal Structure of the World

1 | Scientific Explanation: Three General Conceptions

MODERN SCIENCE provides us with extensive knowledge about the world in which we live. We know that our universe originated in a 'big bang' several thousand million years ago, and that it has been expanding ever since. We know that there was a fairly severe drought in northeastern Arizona during the last few years of the fourteenth century A.D., and that large settlements were abandoned at that time. We know that Halley's comet moves in a roughly elliptical orbit, and that it will return to perihelion in 1986. We know that—barring nuclear holocaust—the human population of the earth will continue to increase for some time to come. We know that the planet Uranus has rings, that *E. coli* live in the normal human intestinal tract, that copper is an electrical conductor, and that the surface temperature of Venus is high. As these examples show, science provides knowledge of what has happened in the past, what will happen in the future, and what is happening now in regions that we are not observing at the present moment. It encompasses knowledge of both particular facts and general regularities. In none of these instances, of course, is our scientific knowledge to be regarded as certain or incorrigible; nevertheless, the physical, biological, and social sciences furnish impressive bodies of knowledge about what goes on in the world, and we have every reason to believe that these sciences will continue to grow at a prodigious rate.

Such knowledge—valuable as it may be for intellectual or practical purposes—is not fully satisfying. Not only do we desire to know *what* happens; we also want to understand *why*. Moreover, it is widely acknowledged today that science *can* provide explanations of natural phenomena; indeed, to many philosophers and scientists, this is the primary goal of scientific activity. Scientific explanations can be given for such particular occurrences as the appearance of Halley's comet in 1759 or the crash of a DC-10 jet airliner in Chicago in 1979, as well as such general features of the world as the nearly elliptical orbits of planets or the electrical conductivity of copper. The chief aim of this book is to try to discover just what scientific understanding of this sort consists in.

Before undertaking the task at hand, it may be useful to make a remark about the basic strategy. During the last thirty-six years (since 1948), a number of quasi-formal 'models' of scientific explanation have appeared in the philosophical literature, and a good deal of attention has been devoted to the investigation of their formal properties. Quite a few are such familiar friends that philosophers often refer to them merely by their initials. We shall meet many of them in the ensuing chapters: deductive-nomological (D-N), deductive-statistical (D-S), inductive-statistical (I-S), statistical-relevance (S-R), deductive-nomological-probabilistic (D-N-P), expected information (E-I), as well as several others for which such initialized designations have not been generally adopted.

Although I shall discuss in some detail various technical aspects of the models, my initial concern will be with what Carnap (1950, sec. 2) called "clarification of the explicandum." Many philosophical studies, including the one to which this book is devoted, aim at providing reasonably precise explications of fundamental concepts, but unless we take preliminary steps to give some understanding of the concept we are trying to explicate—the explicandum—any attempt to formulate an exact explication is apt to be wide of the mark. I am firmly convinced that such terms as "explanation," "scientific explanation," "scientific understanding," and a host of closely related terms are far from being well understood even in a preliminary and imprecise way. Consequently, in an effort to improve that situation, I shall devote the first chapter to a survey of three general conceptions of scientific explanation that have venerable histories and contemporary importance. In the fourth chapter, these three general conceptions will be reconsidered in detail in the light of certain pervasive features of explanations that make some essential appeal to statistical laws. As a result of these efforts at preliminary clarification of the explicandum, we shall find important guidelines for the evaluation of the various 'models' that have been proposed in recent years.

EXPLANATION VERSUS DESCRIPTION

The idea that there are two kinds of scientific knowledge—knowledge of *what* and knowledge of *why*—is not new. In the *Posterior Analytics* (71b18–25), Aristotle distinguishes syllogisms that provide scientific understanding from those that do not. In *The Art of Thinking* ("Port-Royal Logic," first published in 1662), Antoine Arnauld distinguishes demonstrations that merely *convince* the mind from those that also *enlighten* it. He continues, "Enlightenment ought to be the principal fruit of true knowledge. Our minds are unsatisfied unless they know not only *that* a thing is but *why* it is" (1964, p. 330).

The notion that explanation is a major goal of science—or any goal of science at all—has not been universally adopted. In the 1911 edition of

The Grammar of Science, Karl Pearson spoke for a large group of physical scientists when he said, "Nobody believes that science explains anything; we all look upon it as a shorthand description, as an economy of thought" (1957, p. xi). Around the turn of the century, many physicists and chemists held this view about the nonexplanatory character of science because they were convinced that the existence of such micro-entities as electrons, atoms, and molecules could not be scientifically established; indeed, many believed that knowledge of the microstructure of matter is beyond the scope of science. Some argued that this is not really a limitation on the power of science, for the proper business of science is to enable us to make reliable predictions—or, more accurately, to make inferences from observed facts to unobserved facts—but not to try to explain anything. In its most extreme form—which is, incidentally, the version Pearson espoused—this thesis concerning the nature of science holds that the *sole* purpose of science is to enable us to predict our future experiences.

Several distinct issues are involved here. The first issue has to do with *phenomenalism*. This is the doctrine that the only reality consists of human sensations—that what we normally take to be ordinary physical objects are actually nothing more than complex combinations of sensations. One classic expression of this thesis can be found in Ernst Mach's *The Analysis of Sensations* (1914), and it was embraced by a number of early adherents of logical positivism. According to this view, the aim of science is to discover regular patterns among our sensations that will enable us to predict future sensations. Since, in my opinion, phenomenalism has been effectively discredited on a number of grounds, I do not intend to discuss it further in this book. I shall adopt the standpoint of *physicalism*, which holds that perception (of a fairly direct sort) provides us with reliable—though not incorrigible—knowledge of ordinary middle-sized physical objects and events. We have, I believe, sound reasons for taking such entities as flowers, rocks, and sneezes to be real.

The controversy regarding *instrumentalism* and *theoretical realism* is another issue that arises in this context. Instrumentalism is the doctrine that scientific theories that seem to make reference to unobservable entities are not to be considered literally true; instead, they should be regarded as useful instruments for dealing with the observable events, objects, and properties that we find in our environment. For example, some philosophers and scientists have claimed that although such perceptible entities as thermometers, pressure gauges, and containers of gas are real enough, atoms and molecules, which are too small to be observed, are unreal. In response to the successes of the molecular-kinetic theory of gases in the latter part of the nineteenth century, instrumentalists like Pearson and Mach could reply that atoms and molecules are useful fictions that enable us to make excellent predictions about the behavior of gases. At that time, however, many still insisted that there was no compelling evidence that such entities actually exist. As I shall explain in chapter 8, it seems to me that nineteenth-century physical scientists were justified in taking an agnostic view concerning the nature of microphysical entities, but that decisive evidence for the existence of atoms and molecules emerged early in the twentieth century. For the moment, however, I merely want to call attention to the thesis of instrumentalism.

If one maintains that atoms and molecules are real entities, then it may be plausible to claim that they enable us to explain the behavior of gases. Pressure, for example, is explained on the basis of collisions of small material particles (which obey Newton's laws of motion) with the walls of the container. If, however, atoms and molecules are mere fictions, it does not seem reasonable to suppose that we can *explain* the behavior of a gas by saying that it acts *as if* it were composed of small particles. Since instrumentalists do not have any alternative explanation of such phenomena, they have often taken refuge in the view that providing explanations is no part of the scientific enterprise anyhow.

The main idea of Mach, Pearson, and others concerning "shorthand descriptions" and "economy of thought" may be illustrated by Johann Jakob Balmer's formula for a particular series of lines in the spectrum of hydrogen. In 1885, Balmer published a formula for the wavelengths of the spectral lines, which can be written,

$$1/\lambda = R(1/2^2 - 1/n^2),$$

where n takes successive integral values beginning with 3, and R is now known as the Rydberg constant. When he worked out this relationship, Balmer was aware of the wavelengths of four lines that had been measured precisely:

$H_\alpha = 6562.10$ Å ($n = 3$, red)
$H_\beta = 4860.74$ Å ($n = 4$, green)
$H_\gamma = 4340.1$ Å ($n = 5$, blue)
$H_\delta = 4101.2$ Å ($n = 6$, violet)

Additional lines in the Balmer series have subsequently been discovered, and the measured values of their wavelengths agree well with the values that come from Balmer's formula. Obviously, it is more economical to write the simple formula than to keep a list of the numerical values for the wavelengths of all of the lines. Balmer also speculated that his formula could be generalized in the following way:

$$1/\lambda = R(1/m^2 - 1/n^2),$$

where *m* has been substituted for 2 in the previous formula. Additional series that fit this generalized formula have been found:

$m = 1 \rightarrow$ Lyman series
$m = 2 \rightarrow$ Balmer series
$m = 3 \rightarrow$ Paschen series
$m = 4 \rightarrow$ Brackett series
$m = 5 \rightarrow$ Pfund series

This formula yields an even more economical description of the lines in the hydrogen spectrum, but manifestly neither formula *explains* anything about the spectral lines. According to the instrumentalist, it seems, the ultimate aim of science is to find formulas of the foregoing sorts that enable us to describe concisely certain known facts and to predict accurately others that may be discovered in the future.

The instrumentalist view has also been highly influential in psychology, where behaviorists have maintained that their science is concerned only with the stimuli that impinge upon their subjects (humans or animals of other species) and the physical responses that are elicited under specifiable circumstances. These stimuli and responses can be objectively observed and measured. Internal psychological states—for example, feelings of anxiety, hunger, or love—are highly subjective; they are not amenable to objective measurement. They can, however, be regarded as fictions—sometimes called "intervening variables"—which do not literally describe any psychological reality, but which do have instrumental value in describing stimulus-response relationships and in predicting observable behavior. This kind of instrumentalism has remained influential in psychology well into the twentieth century. A classic account of the situation in psychology at mid-century can be found in (MacCorquodale and Meehl, 1948).

In his preface to the third edition of *The Grammar of Science*, just prior to the previously quoted statement about the nonexplanatory character of science, Pearson remarks, "Reading the book again after many years [the first edition was published in 1892, the second in 1900], it is surprising to find how the heterodoxy of the 'eighties had become the commonplace and accepted doctrine of to-day." For some years now, it seems to me, the "commonplace and accepted doctrine" of 1911 has been heterodox, and theoretical realism has become the orthodox view. However, theoretical realism has quite recently come under strong attack by Bas van Fraassen (1980) and Hilary Putnam (1982, 1982a), among others. One wonders to what extent such changes in philosophic doctrine are mere matters of changing fashion, rather than solid results based upon strong philosophical arguments. I shall return to this question in chapter 8.

Although the view that explanation is outside of the domain of science often goes hand in hand with the denial of theoretical realism, they do not always go together. For example, a letter to the editor of the *Scientific American* offered the following comment on an article that had claimed great explanatory merit for the quark hypothesis:

> The quark hypothesis describes the behavior of the subatomic particles in a successful and concise manner. No 'explanation' of these particles' behavior is, of course, ever possible through either pure or experimental science. "Why?" is a question left to philosophy and is not a proper part of any physical theory. (McCalla, 1976, p. 8)

The author of this letter does not appear to be denying the reality either of quarks or of other microphysical entities. It would be incorrect to give the impression that a denial of realism is the only ground on which people maintain that explanation is outside of the domain of science. In some cases, I suspect, such views rest upon the anthropomorphic notion that 'explanations' must appeal to human or superhuman purposes. I do not know what reasons motivated the author of the foregoing letter.

Conversely, it would also be a mistake to suppose that everyone who rejects theoretical realism denies that science has explanatory power. As we shall see in greater detail in chapter 4, van Fraassen, whose agnostic attitude regarding microphysical entities makes him an antirealist, offers a powerful account of scientific explanation (1980, chap. 5).

The question whether science can provide explanations as well as descriptive knowledge arises poignantly in contemporary quantum mechanics. In a popular article entitled "The Quantum Theory and Reality," Bernard d'Espagnat (1979) brings out this issue with admirable clarity. At the outset, he focuses explicit attention upon the problem of explanation:

> Any successful theory in the physical sciences is expected to make accurate predictions. . . . From this point of view quantum mechanics must be judged highly successful. As the fundamental modern theory of atoms, of molecules, of elementary particles, of electromagnetic radiation, and of the solid state it supplies methods for calculating the results of experiments in all these realms.
>
> Apart from experimental confirmation, however, something more is generally demanded of a theory. It is expected not only to determine the results of an experiment but also to provide some understanding of the physical events that are presumed to underlie the observed results. In other words, the theory should not only give the position of a pointer on a dial but also explain why the pointer takes up that position. When one seeks information of this kind in the quantum theory, certain conceptual difficulties arise. (P. 158)

The bulk of the article is occupied with a discussion of some perplexing experiments that are closely related to the famous problem raised by Einstein, Podolsky, and Rosen (1935). Given the recalcitrant character of that problem, the last sentence in the d'Espagnat quotation may just be the understatement of the century. After presenting these problems in some detail, d'Espagnat then considers the possibility of simply denying that science provides explanations, and rejects it. Having noted that the results of the experiments in question agree with the predictions of quantum theory, he continues:

> One conceivable response to the . . . experiments is that their outcome is inconsequential. . . . the results are merely what was expected. They show that the theory is in agreement with experiment and so provide no new information. Such a reaction would be highly superficial. (P. 181)

In this judgment, I believe, d'Espagnat is altogether correct. I shall say more about these issues in chapter 9.

It is now fashionable to say that science aims not merely at describing the world; it also provides *understanding*, *comprehension*, and *enlightenment*. Science presumably accomplishes such high-sounding goals by supplying scientific explanations. The current attitude leaves us with a deep and perplexing question, namely, if explanation does involve something over and above mere description, just what sort of thing is it? The use of such honorific near-synonyms as "understanding," "comprehension," and "enlightenment" makes it sound important and desirable, but does not help at all in the philosophical analysis of explanation—scientific or other. What, over and above descriptive knowledge of the world, is required in order to achieve understanding? This seems to me to pose a serious philosophical problem, especially for those who hold what must be considered the most influential contemporary theory concerning the nature of scientific explanation (see Salmon, 1978).

The main purpose of the present book is to examine the nature of scientific explanation. It will become clear, as the discussion unfolds, that the issues to be considered apply not only to modern physics but rather to the whole range of scientific disciplines from archaeology to zoology, including a great many that occupy alphabetically intermediate positions. The approach will be both iconoclastic and constructive. I shall level what seem to me to be grave criticisms against what currently qualifies as 'the received view,' but I shall also try to elaborate a more satisfactory alternative account.

OTHER TYPES OF EXPLANATION

It is advisable, I believe, to begin with some brief remarks circumscribing the concept that is the object of our attention. The term "explanation" has a number of uses, many of which are beyond the scope of the present discussion. We are often asked, for example, to explain the meaning of a word; in such cases, something akin to a dictionary definition may constitute an adequate response. Someone might ask for an explanation of the meaning of a story or a metaphor; here a translation of figurative or metaphorical language into literal terms seems called for. A friend might ask us please to explain how to find the location of a party; in this instance, a set of detailed instructions or a map might do quite well. In none of these cases is a scientific explanation of a natural phenomenon being requested. It is crucially important to distinguish *scientific* explanations from such other types. Michael Scriven once complained that one of Carl G. Hempel's models could not accommodate the case in which one 'explains' by gestures to a Yugoslav garage mechanic what is wrong with a car. I am in complete agreement with Hempel's response: "This is like objecting to a metamathematical definition of proof on the ground that it does not fit the use of the word 'proof' in 'the proof of the pudding is in the eating,' nor in '86 proof Scotch' " (1965, p. 413). As Hempel remarks, a model of *scientific* explanation that did accommodate such cases would ipso facto be defective.

It is worth noting that none of the foregoing requests for explanations would normally be posed by asking a why-question. In some of the cases, we are asking *what* something means, or *what* is wrong with the car. In other cases, we are asking *how* a mathematical proof goes, or *how* to get to the party. A request for a scientific explanation, in contrast, can always be reasonably posed by means of a why-question. If the request is not originally formulated in such terms, it can, I believe, be recast as a why-question without distortion of meaning (see Bromberger, 1966). Indeed, van Fraassen's recent account of scientific explanations characterizes them *essentially* as answers to why-questions (1980, chap. 5).

It is crucial to recognize, however, that not all—or even most—why-questions are requests for scientific explanations. Why, we might ask, did one employee receive a larger raise than another? Because she had been paid less than a male colleague for doing the same kind of job. In this case, a *moral or legal justification* is the appropriate response. Why, someone might ask, did you go to the drugstore today? The answer, "To get some aspirin for my headache," constitutes a *practical justification*. When Job asked God why he had been singled out for such extraordinary misfortune and suffering, he seems to have been seeking *religious consolation*. It would, of course, spawn endless philosophical mischief to confuse justification or consolation with scientific explanation. Moreover, it would be a sadistic joke to offer a scientific explanation when consolation is sought. To tell a bereaved widow, who asks why her husband was taken from her, that (as she is fully aware) he was decapitated in an automobile accident and that decapitation is invariably fatal, would simply be wanton cruelty.

LAPLACIAN EXPLANATION

Although one particular philosophical account of scientific explanation has enjoyed considerable recent influence, there are others that have both historical and contemporary significance. As a point of departure for discussion of various general conceptions of scientific explanation, I should like to consider a historical example. Father Eusebio Francisco Kino was a Jesuit missionary who, in the latter part of the seventeenth century, founded a series of missions in what is now northern Mexico and southern Arizona. Kino was a highly educated man who was well trained in astronomy, cartography, and mathematics. While in Spain, just prior to his journey to North America, he observed the great comet of 1680. Writing to a colleague, he began by giving a scientifically accurate description of its appearance and motion; he then went on to comment upon its portent for humanity:

> It appears that this comet, which is so large that I do not know whether or not the world has ever seen one like it or so vast, promises, signifies, and threatens many fatalities . . . its influence will not be favorable. And therefore it indicates many calamaties for all Europe . . . and signifies many droughts, hunger, tempests, some earthquakes, great disorders for the human body, discords, wars, many epidemics, fevers, pests, and the deaths of a great many people, especially of some very prominent persons. May God our Lord look upon us with eyes filled with pity.
>
> And because this comet is so large it signifies that its fatalities will be more universal and involve more peoples, persons, and countries. And since it is lasting so long a time . . . it indicates that its evil influence will afflict mortals for many years. (Bolton, 1960, pp. 62–63)

In those days, even the well-educated saw comets as divine omens of disaster. Kino was no ignorant, country parish priest.

Just a few years later, in 1687, Isaac Newton published the *Principia*. In this work, comets were explained as planetlike objects that move around the sun in highly eccentric orbits. The astronomer Edmund Halley—for whom the great comet of 1682 was named, and who was instrumental in securing the publication of the *Principia*—composed an ''Ode to Newton,'' which was prefixed to that work. In it he wrote:

> Now we know
> The sharply veering ways of comets, once
> A source of dread, nor longer do we quail
> Beneath appearances of bearded stars.
>   (Newton, 1947, p. xiv)

The return of Halley's comet in 1759 marked a major triumph of classical mechanics.

Writing at the beginning of the nineteenth century, P. S. Laplace, the famous advocate of mechanistic determinism, drew an eloquent contrast between the scientific understanding of comets provided by Newtonian physics and the superstitious lack of understanding that led people, in pre-Newtonian times, to regard them as mysterious signs sent by inscrutable supernatural powers.

> But as these phenomena occurring and disappearing at long intervals, seemed to oppose the order of nature, it was supposed that Heaven, irritated by the crimes of earth, had created them to announce its vengeance. Thus, the long tail of the comet of 1456 spread terror throughout Europe. . . . This star after four revolutions has excited among us a very different interest. The knowledge of the laws of the system of the world acquired in the interval had dissipated the fears begotten by the ignorance of the true relationship of man to the universe; and Halley, having recognized the identity of this comet with those of 1531, 1607, and 1682, announced its next return for the end of the year 1758 or the beginning of the year 1759. The learned world awaited with impatience this return which was to confirm one of the greatest discoveries that have been made in the sciences. (Laplace, 1951, p. 5)

Although the elimination of superstitious fears is, undoubtedly, one of the major benefits of scientific explanations—and I think it is important to keep that fact in mind at times, such as the present, when irrationalism seems rampant[1]—it can hardly be regarded as their main purpose. For suppose that comets had been considered good omens—much to be hoped for, and sources of great joy when they appeared—but equally haphazard as far as anyone could tell. As objects of rejoicing rather than of fear, they would not have been any less mysterious.

There is some temptation, I believe, to view scientific explanation solely in psychological terms. We begin to wonder about some phenomenon—

---

[1] As just one indication, consider the remarkable strength of the movement to mandate the teaching, in public schools, of the biblical account of the origin of the world, and of life within it, under the title "creation-science." See (Martin Gardner, 1981) for many additional examples, and (Singer and Benassi, 1981) for an interesting discussion.

be it familiar or unfamiliar, an object of joy or an object of fear—and we are intellectually (and perhaps emotionally) ill at ease until we can summon a set of facts that we take to explain it. Whether we have successfully explained the phenomenon, on this view, depends entirely upon whether we have overcome our psychological uneasiness and can feel comfortable with the phenomenon that originally set us to wondering. We need not object to this conception merely on the ground that people often invoke false beliefs and feel comfortable with the 'explanation' thus provided— as when the behavior of a member of a racial minority is 'explained' in terms of a racial stereotype. We can, quite consistently with this approach, insist that adequate explanations must rest upon *true* explanatory bases. Nor need we object on the ground that supernatural 'explanations' are often psychologically appealing. Again, we can insist that the explanation be grounded in *scientific* fact. Even with those restrictions, however, the view that scientific explanation consists in release from psychological uneasiness is unacceptable for two reasons. First, we must surely require that there be some sort of *objective* relationship between the explanatory facts and the fact-to-be-explained. Even if a person were perfectly content with an 'explanation' of the occurrence of storms in terms of falling barometric readings, we should still say that the behavior of the barometer fails objectively to explain such facts. We must, instead, appeal to meteorological conditions. Second, not only is there the danger that people will feel satisfied with scientifically defective explanations; there is also the risk that they will be unsatisfied with legitimate scientific explanations. A yearning for anthropomorphic explanations of all kinds of natural phenomena—for example, the demand that every explanation involve conscious purposes—sometimes leads people to conclude that physics doesn't *really* explain anything at all (recall the letter about the quark theory that was previously quoted in part). Some people have rejected explanations furnished by general relativity on the ground that they cannot visualize a curved four-dimensional space–time. The psychological interpretation of scientific explanation is patently inadequate.

Another conception, equally inadequate, may be suggested by Laplace's remarks. It is sometimes claimed that scientific explanation consists in reducing the unfamiliar to the familiar—as when Newton explained comets by showing that they are objects that behave in essentially the same way as planets, whose motions were by then quite well known. Olbers' paradox constitutes a clear refutation of that thesis. In 1826, the astronomer Heinrich Olbers showed that, on Newtonian principles, the entire night sky should be brilliantly luminous—every region shining at least as brightly as the midday sun.[2] Yet the fact is that for the most part, except for the moon

---

[2] Edmund Halley anticipated Olbers when he wrote (1720), "If the number of the Fixt

and a few points of light coming from visible stars or planets, the night sky is dark. The explanation of this familiar fact, it turns out, takes us into some of the more esoteric reaches of modern cosmology, involving such concepts as the mean free path of photons and non-Euclidean geometry. In this case, the familiar is explained in terms of highly unfamiliar notions. Other examples, which exhibit a similar character, are easily found. It is a familiar fact that a metal table will support a coffee cup, but the explanation of this fact involves the details of atomic structure and the appeal to conceptions as unfamiliar to everyday experience as the Pauli exclusion principle. We need not linger longer over this conception, even though it has had prominent adherents.[3]

Still another characterization of scientific explanation is suggested by Laplace's discussion of comets, namely, that explanation consists in showing that what appears to be haphazard does actually exhibit some regularity. Although this appeal to regularities in nature is a step in the right direction, it certainly cannot be the whole story. The basic reason is that *although some regularities have explanatory power, others simply cry out to be explained.* Newton's theories explained comets by bringing them within the scope of the laws of universal gravitation and motion. In the context of classical physics, that sort of explanation is extremely satisfying, as Laplace made clear. At the same time, Newton's theories also explained the tides. The regular ebb and flow of the tides had been known to mariners for centuries prior to Newton; moreover, the correlation between the behavior of the tides and the position and phase of the moon was well known. These are the sorts of regularities that arouse, rather than satisfy, our intellectual curiosity. Without Newtonian physics, the tides were not really understood at all.

Consider another example. Suppose it has been noticed that on days when clothes hung out on the line tend to dry more slowly than usual, airplanes at nearby airports require greater than average distances to get off of the ground. Mr. Smith, whose business involves the use of a small airplane, complains to his wife one evening about the difficulty he had getting airborne that day. Even if both of them are fully aware of the previously mentioned regularity, it could hardly be considered an expla-

Stars were more than finite, the whole superficies of their apparent Sphere [i.e., the sky] would be luminous.'' Quoted in (Misner et al., 1973, p. 756).

[3] For example, Holton and Brush (1973, p. 185) remark, ''Perhaps it is not too frivolous to hold that 'to explain' means to reduce to the familiar, to establish a relationship between what is to be explained and the (correctly or incorrectly) unquestioned preconceptions.'' Bridgman (1928, p. 37) expresses a similar view, ''I believe that examination will show that the essence of an explanation consists in reducing a situation to elements with which we are so familiar that we accept them as a matter of course, so that our curiosity rests.''

nation of his difficulty for her to mention that the wash had taken an unusually long time to dry that day.[4]

I am dwelling upon Laplace's conception of scientific explanation not merely to exhibit and refute what I take to be patently mistaken notions, but primarily because he provided an account that was especially appropriate within the context of classical physics. Without making any claims about the historical influence of Laplace, I suspect that the conceptions he expressed have had a large—and not altogether beneficial—effect upon contemporary thought about scientific explanation. In the remainder of this chapter, I should like to support and amplify this general assessment. In so doing, I shall extract three distinct general conceptions of scientific explanation, all of which have historical as well as contemporary significance.

THREE BASIC CONCEPTIONS

Laplace attributed our ability to explain comets to our knowledge of the laws of nature. Twentieth-century philosophers have echoed that view by maintaining that, *with the aid of suitable initial conditions, an event is explained by subsuming it under one or more laws of nature*. If these laws are regarded as deterministic, this formulation becomes hardly more than a translation into more up-to-date and less colorful terminology of Laplace's famous statement:

> Given for one instant an intelligence which could comprehend all of the forces by which nature is animated and the respective situation of the beings who compose it—an intelligence sufficiently vast to submit all these data to analysis—it would embrace in the same formula the movements of the greatest bodies of the universe and those of the lightest atom; for it, nothing would be uncertain and the future, as the past, would be present to its eyes. (1951, p. 4)

Such an intelligence would exemplify the highest degree of scientific knowledge; it would, on Laplace's view, be able to provide a complete scientific explanation of any occurrence whatsoever.

There are, it seems to me, at least three distinct ways in which such Laplacian explanations can be construed. In order to relate them to the modern context, we will need to introduce a bit of technical terminology. It is customary, nowadays, to refer to the event-to-be-explained as the *explanandum-event*, and to the statement that such an event has occurred

---

[4] Those readers who are unacquainted with this example can find an explanation of the regularity in chapter 9, pp. 268–69.

as the *explanandum-statement*. Those facts—both particular and general—that are invoked to provide the explanation are known as the *explanans*. If we want to refer specifically to statements that express such facts, we may speak of the *explanans-statements*. The explanans and the explanandum taken together constitute the explanation. Let us now look at the three conceptions.

(1) *Epistemic conception*. Suppose that we attempt to explain some occurrence, such as the appearance of a particular comet at a particular place and time. By citing certain laws, together with suitable initial conditions, we can deduce the occurrence of the event-to-be-explained. By employing observational data collected when his comet appeared in 1682, Halley predicted its return in 1759.[5] These data, along with the laws employed in the deduction, subsequently provided an explanation of that appearance of the comet. This explanation could be described as *an argument to the effect that the event-to-be-explained was to be expected by virtue of the explanatory facts*. The key to this sort of explanation is *nomic expectability*. An event that is quite unexpected in the absence of knowledge of the explanatory facts is rendered expectable on the basis of lawful connections with other facts. Nomic expectability as thus characterized is clearly an epistemological concept. On this view, we can say that there is a relation of *logical necessity* between the laws and the initial conditions on the one hand, and the event-to-be-explained on the other—though it would be more accurate to say that the relation of logical necessity holds between the explanans-statements and the explanandum-statement.

(2) *Modal conception*. Under the same circumstances we can say, alternatively, that because of the lawful relations between the antecedent conditions and the event-to-be-explained there is a relation of *nomological necessity* between them. In Laplace's *Essay*, the discussion of determinism is introduced by the following remarks:

> All events, even those which on account of their insignificance do not seem to follow the great laws of nature, are a result of it just as necessarily as the revolutions of the sun. In ignorance of the ties which unite such events to the entire system of the universe, they have been made to depend upon final causes or upon hazard . . . but these imaginary causes have gradually receded with the widening bounds of knowledge and disappear entirely before sound philosophy, which sees in them only the expression of our ignorance of the true causes. (1951, p. 3)

[5] Actually Halley did not make a very precise prediction, for he did not take account of the perturbations in the orbit due to Jupiter and Saturn. This was done by Clairaut; see (Laplace, 1951, p. 6).

Nomological necessity, it might be said, derives from the laws of nature in much the same way as logical necessity rests upon the laws of logic. *In the absence of knowledge of the explanatory facts, the explanandum-event* (the appearance of the comet) was something that *might not have occurred for all we know; given the explanatory facts it had to occur*. The explanation exhibits the nomological necessity of the fact-to-be-explained, given the explanatory facts. Viewing the matter in this way, one need not maintain that an explanation is an argument showing that the explanandum-event had to occur, given the initial conditions. Although a deductive argument can be constructed (as in the epistemic account) within which a relation of logical entailment obtains, an explanation need not be regarded as such an argument, or as any kind of argument at all. In comparing the epistemic and modal conceptions, it is important to be clear on the roles of the two kinds of necessity. In the epistemic conception, the relation of *logical* necessity obtains between the entire explanans and the explanandum by virtue of the laws of deductive logic. In the modal conception, the relation of *physical* necessity holds between particular antecedent conditions and the explanandum-event by virtue of the general laws, which we are taking to be part of the explanans.[6]

(3) *Ontic conception.* There is still another way of looking at Laplacian explanations. If the universe is, in fact, deterministic, then nature is governed by strict laws that *constitute* natural regularities. Law-statements describe these regularities. Such regularities endow the world with patterns that can be discovered by scientific investigation, and that can be exploited for purposes of scientific explanation. To explain an event—to relate the event-to-be-explained to some antecedent conditions by means of laws—is to fit the explanandum-event into a discernible pattern. This view seems to be present in Laplace's thought, for he remarks that comets "seemed

[6] The contrast being suggested is well illustrated by a controversy between Hempel and Scriven concerning the role of laws in scientific explanation. As we have seen, Hempel (1965) insists that general laws be present in the explanans. Scriven (though he is not a proponent of the modal conception) argues that a set of particular antecedent conditions may constitute an adequate explanation of a particular event; consequently, the explanans need not include reference to any general laws. A law that provides a connection between the explanans and the explanandum constitutes a "role-justifying ground" for the explanation by showing, roughly speaking, that the explanans is explanatorily relevant to the explanandum. For Hempel, the laws of logic—which provide the relation of relevance of the explanans to the explanandum—are not part of the explanation, but can be called upon to justify the claim that a given explanans has explanatory force with respect to some explanandum. Scriven invokes similar considerations to argue that general laws of nature should remain outside of scientific explanations to be called upon, if necessary, to support the claim that a given explanation is adequate. See (Scriven, 1959) for details; Hempel's reply is given in his (1965, pp. 359–364).

to oppose the order of nature" before we knew how to explain them, but that subsequent "knowledge of the laws of the system of the world" provided understanding of them (1951, p. 5). Moreover, as noted previously, he speaks of "the ties which unite such events to the entire system of the universe" (1951, p. 3). Because of the universal (nonstatistical) character of the laws involved in Laplacian explanations, we can also say that given certain portions of the pattern of events, and given the lawful relations to which the constituents of the patterns conform, other portions of the pattern of events must have certain characteristics. Looking at explanation in this way, we might say that *to explain an event is to exhibit it as occupying its* (nomologically necessary) *place in the discernible patterns of the world.*

These three general conceptions of scientific explanation all seem to go back at least to Aristotle. We have already remarked on his identification of certain sorts of syllogisms as explanations; this conforms to the epistemic conception that regards explanations as deductive arguments.[7] He seems to be expressing the modal conception when he remarks that "the proper object of unqualified scientific knowledge is something which cannot be other than it is" (*Posterior Analytics*, 1. 2. 71b14–16). And in the same context, discussing the nature of the syllogism that yields "scientific knowledge," he says, "The premises must be the causes of the conclusion, better known than it, and prior to it; its causes, since we possess scientific knowledge of a thing only when we know its cause; prior, in order to be causes; antecedently known, this antecedent knowledge being not our mere understanding of the meaning, but knowledge of the fact as well" (ibid., 71b29–33). These remarks suggest an ontic conception.

In the twentieth century, we still find the same three notions figuring prominently in philosophical discussions of scientific explanation. The *epistemic conception* represents the currently 'received view,' which has been advocated by such influential philosophers as Braithwaite, Hempel, Nagel, and Popper. It was succinctly formulated by Hempel in the following way:

[7] In the *Posterior Analytics* (1928, 1. 2. 71b18–24), Aristotle writes: "By demonstration I mean a syllogism productive of scientific knowledge, a syllogism, that is, the grasp of which is *eo ipso* such knowledge. Assuming that my thesis as to the nature of scientific knowledge is correct, the premises of demonstrated knowledge must be true, primary, immediate, better known than and prior to the conclusion, which is further related to them as effect to cause. Unless these conditions are satisfied, the basic truths will not be 'appropriate' to the conclusion. Syllogism there may indeed be without these conditions, but such syllogism, not being productive of scientific knowledge, will not be demonstration." Richard Jeffrey (1969) offers an illuminating comparison between this Aristotelian view and Hempel's D-N account.

[An] explanatory account may be regarded as an argument to the effect that the event to be explained . . . was to be expected by reason of certain explanatory facts. These may be divided into two groups: (i) particular facts and (ii) uniformities expressed by general laws. (1962a, p. 10)

The *modal conception* has been clearly affirmed by D. H. Mellor: "The thesis is that we call for explanation only of what, although we know it is so, might have been otherwise for all else of some suitable sort we know" (1976, p. 234). In what does an explanation consist?

We want to know why what might not have happened nonetheless did. Causal explanation closes the gap by deducing what happened from known earlier events and deterministic laws. So in this respect it satisfies the demand for explanation: what follows from what is true must also be true. Given the causal explanans, things *could not have happened otherwise* than the explanandum says. (1976, p. 235, italics added)

G. H. von Wright (1971) gives concise expression to this same conception: "What makes a deductive-nomological explanation 'explain,' is, one might say, that it tells us why *E had* to be (occur), why *E* was *necessary* once the basis [body of explanatory facts] is there and the laws are accepted" (p. 13, italics in original). This same view can be found explicitly in C. S. Peirce (1932, 2:776).

The *ontic conception* is the one for which I shall be arguing. In Salmon (1977a, p. 162), I offered the following characterization: "To give scientific explanations is to show how events . . . fit into the causal structure of the world." Hempel summarizes the import of his major monographic essay, "Aspects of Scientific Explanation" (1965a, p. 488), in rather similar terms: "The central theme of this essay has been, briefly, that all scientific explanation involves, explicitly or by implication, a subsumption of its subject matter under general regularities; that it seeks to provide a systematic understanding of empirical phenomena by showing that they fit into a nomic nexus." I find this statement by Hempel in almost complete accord with the viewpoint I shall be advocating; my suggestion for modification would be to substitute the words "*how* they fit into a *causal* nexus" for "*that* they fit into a *nomic* nexus." It seems to me that Hempel began the "Aspects" article with statements clearly indicating that he embraced the epistemic conception, but he ended with a summary that seems closer to the ontic conception. Because these three conceptions had not been explicitly formulated and distinguished at the time of his writing, he was, I think, unaware of any conflict. As we shall see in subsequent chapters, there are profound differences, especially in the context of statistical explanation.

Those philosophers who have adopted the ontic conception of scientific explanation have generally regarded the pattern into which events are to be fit as a causal pattern. This feature of the view is brought out explicitly in the quotation from Aristotle. It certainly was present in Laplace's mind when he wrote, "Present events are connected with preceding ones by a tie based upon the evident principle that a thing cannot occur without a cause which produces it" (1951, p. 3), and "We ought then to regard the present state of the universe as the effect of its anterior state and as the cause of the one which is to follow" (1951, p. 4). It was also explicit in my formulation quoted previously. Hempel, however, does not share this notion; for him the pattern is lawful (nomic), but the laws involved need not be causal laws (1965, pp. 352–354). In view of well-known Humean problems associated with causality, it might *seem* desirable to try to avoid reference to causal laws in dealing with scientific explanation. Nevertheless, I shall try to show that we need not purge the causal notions; indeed, I shall argue that they are required for an adequate theory of scientific explanation. In order to implement the causal version of the ontic conception, however, it will be necessary to examine the nature of causal relations with considerable care, and to show how they can be employed unobjectionably in a theory of scientific explanation. This problem will be postponed until chapters 5–7.

The foregoing three ways of thinking about scientific explanation may *seem* more or less equivalent—with somewhat distinct emphases perhaps—but hardly more than different verbal formulations. This is true as long as we are talking about the kind of explanation that involves appeal to universal laws only. A striking divergence will appear, however, when we consider explanations that invoke statistical (or probabilistic) laws. In the deterministic framework of Laplace's thought, all of the fundamental laws of nature are taken to be strictly universal; any appeal to probabilities is merely a reflection of human ignorance. In twentieth-century science, the situation is quite different. There is a strong presumption in contemporary physics that some of the basic laws of nature may be irreducibly statistical—that probability relations *may* constitute a fundamental feature of the physical world. There is, to be sure, some disagreement as to whether determinism is true or false—whether modern physics requires an indeterministic interpretation. I do not want to prejudge this issue. In the attempt to elaborate a philosophical theory of scientific explanation, it seems to me, we must try to construct one that will be viable in either case. Therefore, we must leave open the possibility that some scientific explanations will be unavoidably statistical. This means that we must pay careful attention to the nature of statistical explanation.

An Outline of Strategy

Much of the contemporary literature on scientific explanation arises directly or indirectly in response to the classic 1948 Hempel-Oppenheim paper, "Studies in the Logic of Explanation."[8] In it the authors attempt to provide a precise explication of what has come to be known as the deductive-nomological or D-N model of scientific explanation. They did not invent this mode of scientific explanation, nor were they the first philosophers to attempt to characterize it. As mentioned previously, its roots go back at least to Aristotle, and it is strongly suggested in such works as Arnauld's *The Art of Thinking* ("Port-Royal Logic") and Laplace's *Philosophical Essay on Probabilities*. In none of the anticipations by these or other authors, however, do we find the precision and detail of the Hempel-Oppenheim account. One might almost say that 1948 marks the division between the prehistory and the history of the philosophical study of scientific explanation. When other such influential philosophers as R. B. Braithwaite (1953), Ernest Nagel (1961), and Karl R. Popper (1935, 1959) espoused a similar account of deductive explanation, it achieved virtually the status of a 'received view.'[9]

According to the 'received view,' particular facts are explained by subsuming them under general laws, while general regularities are explained by subsumption under still broader laws. If a particular fact is successfully subsumed under a lawful generalization, it is, on this view, completely explained. One can legitimately ask for an explanation of the general law that figures in the explanation, but an explanation of the general law would be a different and additional explanation, not an essential part of the original explanation of the particular fact. For example, to explain why this particular penny conducts electricity, it suffices to point out that it is composed of copper and that all copper objects conduct electricity. If we are asked to explain why copper conducts electricity, we may give a further *distinct* explanation in terms of the fact that copper is a metal with conduction electrons that are not tightly bound to individual atoms and are free to move when an electric potential is applied.

Most proponents of this subsumption theory maintain that some events can be explained statistically by subsumption under statistical laws in much the same way that other events—such as the fact that the penny just inserted

---

[8] See (Rescher, 1970) for an extensive bibliography on scientific explanation up to the date of its publication.

[9] Although Popper's *Logik der Forschung* (1935) contains an important anticipation of the D-N model, it does not provide as precise an analysis as was embodied in (Hempel and Oppenheim, 1948). Moreover, Popper's views on scientific explanation were not widely influential until the English translation (Popper, 1959) of his 1935 book appeared. It is for these reasons that I chose 1948, rather than 1935, as the critical point of division between the history and the prehistory of the subject.

behind the fuse conducts electricity—are explained by appeal to universal laws. Thus we can explain the fact that a particular window was broken by pointing out that it was struck by a flying baseball, even though not all, but only most, windows so struck will shatter.

Although I disagreed from the beginning with the proponents of the *standard* subsumption view about the nature of the relation of subsumption of particular facts under universal or statistical generalizations, I did for some time accept the notion that *suitable* subsumption under generalizations is sufficient to explain particular facts. In *Statistical Explanation and Statistical Relevance* (Salmon et al., 1971), I tried to give a detailed account of what seemed to me the appropriate way to subsume facts under general laws for purposes of explanation. This effort led to the elaboration of the statistical-relevance (S-R) model of scientific explanation. As the name suggests, statistical relevance relations play a key role in this model of scientific explanation.

Subsequent reflection has convinced me that subsumption of the foregoing sort is only part—not all—of what is involved in the explanation of particular facts. It now seems to me that explanation is a two-tiered affair. At the most basic level, it is necessary, for purposes of explanation, to subsume the event-to-be-explained under an appropriate set of statistical relevance relations, much as was required under the S-R model. At the second level, it seems to me, the statistical relevance relations that are invoked at the first level must be explained in terms of *causal* relations. The explanation, on this view, is incomplete until the causal components of the second level have been provided. This constitutes a sharp divergence from the approach of Hempel, who explicitly rejects the demand for causal laws (1965, pp. 352–354).

It would be advisable, I believe, to adopt an approach similar to one suggested by Wolfgang Stegmüller (1973, p. 345), who characterized the kind of subsumption under statistical relevance relations provided by the S-R model as "statistical analysis" rather than "statistical explanation." The latter term is reserved for the entity that comprises both the statistical-relevance level and the causal level as well. As (Humphreys, 1981, 1983) and (Rogers, 1981) persuasively argue, statistical analyses have important uses, but they fall short of providing genuine scientific understanding. To emphasize this point, I shall use the term *S-R basis* to refer to the statistical component of an explanation.[10]

The remainder of the present book is divided into two main parts,

---

[10] In relinquishing the thesis that the S-R model provides an adequate characterization of scientific explanation, I accept as valid most of the criticisms leveled against it by Achinstein (1983). These criticisms do not, however, undermine the utility of the S-R basis as a foundation for scientific explanations.

corresponding to the two tiers just mentioned. Chapters 2–3, which constitute the first part, deal essentially with the S-R basis; it is regarded as an indispensable portion of the present account of scientific explanation, even though it is no longer taken as a complete model. These two chapters contain a number of important revisions of the S-R model itself, vis-à-vis previous presentations, as well as a good deal of supplementary material. Chapter 3 deals with the difficult concept of physical randomness—that is, objective homogeneity of reference classes—which is required to implement an ontic treatment of statistical explanation adequate to the possibly indeterministic context of contemporary science. The result, I hope, is a substantially improved version of the S-R model—one that can provide a satisfactory basis for the second level. These two chapters contain, roughly, all that can be said, as far as I am aware, regarding scientific explanation of particular facts without invoking causal considerations.

Chapter 4 is a transitional discussion of the three general conceptions (introduced in this chapter) in the light of statistical considerations. It constitutes a serious attempt to provide a much-needed clarification of the explicandum. The conclusion drawn from the discussion is that the ontic conception is the only acceptable one.

Chapters 5–8, which make up the second main part, deal explicitly with causality in scientific explanation. In order to achieve the goal of explicating the role of causality in scientific explanation, it is necessary to develop a theory of causality that, though it borrows heavily from other authors, incorporates various novel elements. Among its conspicuous (though not necessarily novel) features are (1) it is a probabilistic or statistical concept, (2) it places great emphasis upon the distinction between causal *processes* and causal *interactions*, and (3) it takes processes to be more fundamental than events. The reader will have to judge for herself or himself concerning the adequacy of the account of causality offered in these chapters, and its fertility in providing an improved theory of scientific explanation. It is my hope that the result is a theory of scientific explanation that constitutes a significant advance beyond its predecessors.

Chapter 9, the concluding chapter, deals with some general features of my approach to scientific explanation, including some of its shortcomings and limitations. It points the way, I believe, to the kind of research that should extend and deepen our philosophical understanding of scientific explanation.

2 | Statistical Explanation and Its Models

THE PHILOSOPHICAL THEORY of scientific explanation first entered the twentieth century in 1962, for that was the year of publication of the earliest bona fide attempt to provide a systematic account of statistical explanation in science.[1] Although the need for some sort of inductive or statistical form of explanation had been acknowledged earlier, Hempel's essay "Deductive-Nomological vs. Statistical Explanation" (1962) contained the first sustained and detailed effort to provide a precise account of this mode of scientific explanation. Given the pervasiveness of statistics in virtually every branch of contemporary science, the late arrival of statistical explanation in philosophy of science is remarkable. Hempel's initial treatment of statistical explanation had various defects, some of which he attempted to rectify in his comprehensive essay "Aspects of Scientific Explanation" (1965a). Nevertheless, the earlier article did show unmistakably that the construction of an adequate model for statistical explanation involves many complications and subtleties that may have been largely unanticipated. Hempel never held the view—expressed by some of the more avid devotees of the D-N model—that *all* adequate scientific explanations must conform to the deductive-nomological pattern. The 1948 Hempel-Oppenheim paper explicitly notes the need for an inductive or statistical model of scientific explanation in order to account for some types of legitimate explanation that actually occur in the various sciences (Hempel, 1965, pp. 250–251). The task of carrying out the construction was, however, left for another occasion. Similar passing remarks about the need for inductive or statistical accounts were made by other authors as well, but the project was not undertaken in earnest until 1962—a striking delay of fourteen years after the 1948 essay.

One can easily form the impression that philosophers had genuine feelings of ambivalence about statistical explanation. A vivid example can be

---

[1] Ilkka Niiniluoto (1981, p. 444) suggests that "Peirce should be regarded as the true founder of the theory of inductive-probabilistic explanation" on account of this statement, "The statistical syllogism may be conveniently termed the explanatory syllogism" (Peirce, 1932, 2:716). I am inclined to disagree, for one isolated and unelaborated statement of that sort can hardly be considered even the beginnings of any geniune theory.

found in Carnap's *Philosophical Foundations of Physics* (1966), which was based upon a seminar he offered at UCLA in 1958.[2] Early in the first chapter, he says:

> The general schema involved in *all explanation* can be expressed symbolically as follows:
>
> 1. $(x) (Px \supset Qx)$
> 2. Pa
> 3. Qa
>
> The first statement is the universal law that applies to any object x. The second statement asserts that a particular object a has the property P. These two statements taken together enable us to derive logically the third statement: object a has the property Q. (1966, pp. 7–8, italics added)

After a single intervening paragraph, he continues:

> At times, in giving an explanation, the only *known* laws that apply are statistical rather than universal. In such cases, we must be content with a statistical explanation. (1966, p. 8, italics added)

Farther down on the same page, he assures us that "these are genuine explanations," and on the next page he points out that "In quantum theory . . . we meet with statistical laws that may not be the result of ignorance; they may express the basic structure of the world." I must confess to a reaction of astonishment at being told that all explanations are deductive-nomological, but that some are not, because they are statistical. This lapse was removed from the subsequent paperback edition (Carnap, 1974), which appeared under a new title.

Why did it take philosophers so long to get around to providing a serious treatment of statistical explanation? It certainly was not due to any absence of statistical explanations in science. In antiquity, Lucretius (1951, pp. 66–68) had based his entire cosmology upon explanations involving spontaneous swerving of atoms, and some of his explanations of more restricted phenomena can readily be interpreted as statistical. He asks, for example, why it is that Roman housewives frequently become pregnant after sexual intercourse, while Roman prostitutes to a large extent avoid doing so. Conception occurs, he explains, as a result of a collision between a male seed and a female seed. During intercourse the prostitutes wiggle their

---

[2] As Carnap reports in the preface, the seminar proceedings were recorded and transcribed by his wife. Martin Gardner edited—it would probably be more accurate to say "wrote up"—the proceedings and submitted them to Carnap, who rewrote them extensively. There is little doubt that Carnap saw and approved the passages I have quoted.

hips a great deal, but wives tend to remain passive; as everyone knows, it is much harder to hit a moving target (1951, p. 170).[3] In the medieval period, St. Thomas Aquinas asserted:

> The majority of men follow their passions, which are movements of the sensitive appetite, in which movements of heavenly bodies can cooperate: but few are wise enough to resist these passions. Consequently astrologers are able to foretell the truth in the majority of cases, especially in a general way. But not in particular cases; for nothing prevents man resisting his passions by his free will. (1947, 1:Qu. 115, a. 4, *ad* Obj. 3)

Astrological explanations are, therefore, of the statistical variety. Leibniz, who like Lucretius and Aquinas was concerned about human free will, spoke of causes that incline but do not necessitate (1951, p. 515; 1965, p. 136).

When, in the latter half of the nineteenth century, the kinetic-molecular theory of gases emerged, giving rise to classical statistical mechanics, statistical explanations became firmly entrenched in physics. In this context, it turns out, many phenomena that *for all practical purposes* appear amenable to strict D-N explanation—such as the melting of an ice cube placed in tepid water—must be admitted *strictly speaking* to be explained statistically in terms of probabilities almost indistinguishable from unity. On a smaller scale, Brownian motion involves probabilities that are, both theoretically and practically, definitely less than one. Moreover, two areas of nineteenth-century biology, Darwinian evolution and Mendelian genetics, provide explanations that are basically statistical. In addition, nineteenth-century social scientists approached such topics as suicide, crime, and intelligence by means of "moral statistics" (Hilts, 1973).

In the present century, statistical techniques are used in virtually every branch of science, and we may well suppose that most of these disciplines, if not all, offer statistical explanations of some of the phenomena they treat. The most dramatic example is the statistical interpretation of the equations of quantum mechanics, provided by Max Born and Wolfgang Pauli in 1926–1927; with the aid of this interpretation, quantum theory explains an impressive range of physical facts.[4] What is even more im-

---

[3] Lucretius writes: "A woman makes conception more difficult by offering a mock resistance and accepting Venus with a wriggling body. She diverts the furrow from the straight course of the ploughshare and makes the seed fall wide of the plot. These tricks are employed by prostitutes for their own ends, so that they may not conceive *too frequently* and be laid up by pregnancy" (1951, p. 170, italics added).

[4] See (Wessels, 1982), for an illuminating discussion of the history of the statistical interpretation of quantum mechanics.

portant is that this interpretation brings in statistical considerations at the most basic level. In nineteenth-century science, the use of probability reflected limitations of human knowledge; in quantum mechanics, it looks as if probability may be an ineluctable feature of the physical world. The Nobel laureate physicist Leon Cooper expresses the idea in graphic terms: "Like a mountain range that divides a continent, feeding water to one side or the other, the probability concept is the divide that separates quantum theory from all of physics that preceded it" (1968, p. 492). Yet it was not until 1962 that any philosopher published a serious attempt at characterizing a statistical pattern of scientific explanation.

INDUCTIVE-STATISTICAL EXPLANATION

When it became respectable for empirically minded philosophers to admit that science not only describes and predicts, but also explains, it was natural enough that primary attention should have been directed to classic and beautiful examples of deductive explanation. Once the D-N model had been elaborated, either of two opposing attitudes might have been taken toward inductive or statistical explanation by those who recognized the legitimacy of explanations of this general sort. It might have been felt, on the one hand, that the construction of such a model would be a simple exercise in setting out an analogue to the D-N model or in relaxing the stringent requirements for D-N explanation in some straightforward way. It might have been felt, on the other hand, that the problems in constructing an appropriate inductive or statistical model were so formidable that one simply did not want to undertake the task. Some philosophers may unreflectingly have adopted the former attitude; the latter, it turns out, is closer to the mark.

We should have suspected as much. If D-N explanations are deductive arguments, inductive or statistical explanations are, presumably, inductive arguments. This is precisely the tack Hempel took in constructing his inductive-statistical or I-S model. In providing a D-N explanation of the fact that this penny conducts electricity, one offers an explanans consisting of two premises: the particular premise that this penny is composed of copper, and the universal law-statement that all copper conducts electricity. The explanandum-statement follows deductively. To provide an I-S explanation of the fact that I was tired when I arrived in Melbourne for a visit in 1978, it could be pointed out that I had been traveling by air for more than twenty-four hours (including stopovers at airports), and almost everyone who travels by air for twenty-four hours or more becomes fatigued. The explanandum gets strong inductive support from those premises; the event-to-be-explained is thus subsumed under a statistical generalization.

It has long been known that there are deep and striking disanalogies between inductive and deductive logic.[5] Deductive entailment is transitive; strong inductive support is not. Contraposition is valid for deductive entailments; it does not hold for high probabilities. These are *not* relations that hold in some approximate way if the probabilities involved are high enough; once we abandon strict logical entailment, and turn to probability or inductive support, they break down entirely. But much more crucially, as Hempel brought out clearly in his 1962 essay, the deductive principle that permits the addition of an arbitrary term to the antecedent of an entailment does not carry over at all into inductive logic. If A entails B, then $A.C$ entails B, whatever C may happen to stand for. However, no matter how high the probability of B given A, there is no constraint whatever upon the probability of B given both A and C. To take an extreme case, the probability of a prime number being odd is one, but the probability that a prime number smaller than 3 is odd has the value zero. For those who feel uneasy about applying probability to cases of this arithmetical sort, we can readily supply empirical examples. A thirty-year-old Australian with an advanced case of lung cancer has a low probability of surviving for five more years, even though the probability of surviving to age thirty-five for thirty-year-old Australians in general is quite high. It is *this* basic disanalogy between deductive and inductive (or probabilistic) relations that gives rise to what Hempel called *the ambiguity of inductive-statistical explanation*—a phenomenon that, as he emphasized, has no counterpart in D-N explanation. His *requirement of maximal specificity* was designed expressly to cope with the problem of this ambiguity.

Hempel illustrates the ambiguity of I-S explanation, and the need for the requirement of maximal specificity, by means of the following example (1965, pp. 394–396). John Jones recovers quickly from a streptococcus infection, and when we ask why, we are told that he was given penicillin, and that almost all strep infections clear up quickly after penicillin is administered. The recovery is thus rendered probable relative to these explanatory facts. There are, however, certain strains of streptococcus bacteria that are resistant to penicillin. If, in addition to the above facts, we were told that the infection is of the penicillin-resistant type, then we would have to say that the prompt recovery is rendered *improbable* relative to the available information. It would clearly be scientifically unacceptable to ignore such relevant evidence as the penicillin-resistant character of the

---

[5] These are spelled out in detail in (Salmon, 1965a). See (Salmon, 1967, pp. 109–111) for a discussion of the 'almost-deduction' conception of inductive inference.

infection; the requirement of maximal specificity is designed to block statistical explanations that thus omit relevant facts. It says, in effect, that when the class to which the individual case is referred for explanatory purposes—in this instance, the class of strep infections treated by penicillin—is chosen, we must not know how to divide it into subsets in which the probability of the fact to be explained differs from its probability in the entire class. If it has been ascertained that this particular case involved the penicillin-resistant strain, then the original explanation of the rapid recovery would violate the requirement of maximal specificity, and for that reason would be judged unsatisfactory.[6]

Hempel conceived of D-N explanations as valid deductive arguments satisfying certain additional conditions. Explanations that conform to his inductive-statistical or I-S model are correct inductive arguments also satisfying certain additional restrictions. Explanations of both sorts can be characterized in terms of the following four conditions:

1. The explanation is an argument with correct (deductive or inductive) logical form,
2. At least one of the premises must be a (universal or statistical) law,
3. The premises must be true, and
4. The explanation must satisfy the requirement of maximal specificity.

This fourth condition is automatically satisfied by D-N explanations by virtue of the fact that their explanatory laws are universal generalizations. If all A are B, then obviously there is no subset of A in which the probability of B is other than one. This condition has crucial importance with respect to explanations of the I-S variety. In general, according to Hempel (1962a, p. 10), an explanation is an argument (satisfying these four conditions) to the effect that the event-to-be-explained was to be expected by virtue of certain explanatory facts. In the case of I-S explanations, this means that the premise must lend high inductive probability to the conclusion—that is, the explanandum must be highly probable with respect to the explanans.

Explanations of the D-N and I-S varieties can therefore be schematized as follows (Hempel, 1965, pp. 336, 382):

$$C_1, C_2, \ldots, C_j \quad \text{(particular explanatory conditions)}$$

(D-N)

$$\frac{L_1, L_2, \ldots, L_k}{E} \quad \begin{array}{l} \text{(general laws)} \\ \text{(fact-to-be-explained)} \end{array}$$

[6] We shall see in chapter 3 that the requirement of maximal specificity, as formulated by Hempel in his (1965) and revised in his (1968), does not actually do the job. Nevertheless, this was clearly its intent.

The single line separating the premises from the conclusion signifies that the argument is deductively valid.

$$C_1, C_2, \ldots, C_j \quad \text{(particular explanatory conditions)}$$

(I-S) $\quad L_1, L_2, \ldots, L_k \quad$ (general laws, at least one statistical)
$$\frac{\qquad\qquad}{E} \, [r]$$
$$E \qquad \text{(fact-to-be-explained)}$$

The double lines separating the premises from the conclusion signifies that the argument is inductively correct, and the number r expresses the degree of inductive probability with which the premises support the conclusion. It is presumed that r is fairly close to one.[7]

The high-probability requirement, which seems such a natural analogue of the deductive entailment relation, leads to difficulties in two ways. First, there are arguments that fulfill all of the requirements imposed by the I-S model, but that patently do not constitute satisfactory scientific explanations. One can maintain, for example, that people who have colds will probably get over them within a fortnight if they take vitamin C, but the use of vitamin C may not explain the recovery, since almost all colds clear up within two weeks regardless. In arguing for the use of vitamin C in the prevention and treatment of colds, Linus Pauling (1970) does not base his claims upon the high probability of avoidance or quick recovery; instead, he urges that massive doses of vitamin C have a bearing upon the probability of avoidance or recovery—that is, the use of vitamin C is relevant to the occurrence, duration, and severity of colds. A *high* probability of recovery, given use of vitamin C, does not confer explanatory value upon the use of this drug with respect to recovery. An *enhanced* probability value does indicate that the use of vitamin C may have some explanatory force. This example, along with a host of others which, like it, fulfill all of Hempel's requirements for a correct I-S explanation, shows that fulfilling these requirements does not constitute a sufficient condition for an adequate statistical explanation.

At first blush, it might seem that the type of relevance problem illustrated by the foregoing example was peculiar to the I-S model, but Henry Kyburg (1965) showed that examples can be found which demonstrate that the

---

[7] It should be mentioned in passing that Hempel (1965, pp. 380–381) offers still another model of scientific explanation that he characterizes as deductive-statistical (D-S). In an explanation of this type, a statistical regularity is explained by deducing it from other statistical laws. There is no real need, however, to treat such explanations as a distinct type, for they fall under the D-N schema, just given, provided we allow that at least one of the laws may be statistical. In the present context, we are concerned only with statistical explanations of nondeductive sorts.

D-N model is infected with precisely the same difficulty. Consider a sample of table salt that dissolves upon being placed in water. We ask why it dissolves. Suppose, Kyburg suggests, that someone has cast a dissolving spell upon it—that is, someone wearing a funny hat waves a wand over it and says, "I hereby cast a dissolving spell upon you." We can then 'explain' the phenomenon by mentioning the dissolving spell—without for a moment believing that any actual magic has been accomplished—and by invoking the true universal generalization that all samples of table salt that have been hexed in this manner dissolve when placed in water. Again, an argument that satisfies all of the requirements of Hempel's model patently fails to qualify as a satisfactory scientific explanation because of a failure of relevance. Given Hempel's characterizations of his D-N and I-S models of explanation, it is easy to construct any number of 'explanations' of either type that invoke some irrelevancy as a purported explanatory fact.[8] This result casts serious doubt upon the entire epistemic conception of scientific explanation, as outlined in the previous chapter, insofar as it takes all explanations to be arguments of one sort or another.

The diagnosis of the difficulty can be stated very simply. Hempel's requirement of maximal specificity (RMS) guarantees that *all* known relevant facts must be included in an adequate scientific explanation, but there is no requirement to insure that *only* relevant facts will be included. The foregoing examples bear witness to the need for some requirement of this latter sort. To the best of my knowledge, the advocates of the 'received view' have not, until recently, put forth any such additional condition, nor have they come to terms with counterexamples of these types in any other way.[9] James Fetzer's *requirement of strict maximal specificity*, which rules out the use in explanations of laws that mention nomically irrelevant properties (Fetzer, 1981, pp. 125–126), seems to do the job. In fact, in (Salmon, 1979a, pp. 691–694), I showed how Reichenbach's theory of nomological statements could be used to accomplish the same end.

The second problem that arises out of the high-probability requirement is illustrated by an example furnished by Michael Scriven (1959, p. 480). If someone contracts paresis, the straightforward explanation is that he

---

[8] Many examples are presented and analyzed in (Salmon et al., 1971, pp. 33–40). Nancy Cartwright (1983, pp. 26–27) errs when she attributes to Hempel the requirement that a statistical explanation increase the probability of the explanandum; this thesis, which I first advanced in (Salmon, 1965), was never advocated by Hempel. Shortly thereafter (1983, 28–29), she provides a correct characterization of the relationships among the views of Hempel, Suppes, and me.

[9] In Hempel's most recent discussion of statistical explanation, he appears to maintain the astonishing view that although such examples have *psychologically* misleading features, they do qualify as *logically* satisfactory explanations (1977, pp. 107–111).

was infected with syphilis, which had progressed through the primary, secondary, and latent stages without treatment with penicillin. Paresis is one form of tertiary syphilis, and it never occurs except in syphilitics. Yet far less than half of those victims of untreated latent syphilis ever develop paresis. Untreated latent syphilis is the explanation of paresis, but it does not provide any basis on which to say that the explanandum-event was to be expected by virtue of these explanatory facts. Given a victim of latent untreated syphilis, the odds are that he will *not* develop paresis. Many other examples can be found to illustrate the same point. As I understand it, mushroom poisoning may afflict only a small percentage of individuals who eat a particular type of mushroom (Smith, 1958, Introduction), but the eating of the mushroom would unhesitatingly be offered as the explanation in instances of the illness in question. The point is illustrated by remarks on certain species in a guide for mushroom hunters (Smith, 1958, pp. 34, 185):

> *Helvella infula*, "Poisonous to some, but edible for most people. Not recommended."
> *Chlorophyllum molybdites*, "Poisonous to some but not to others. Those who are not made ill by it consider it a fine mushroom. The others suffer acutely."

These examples show that high probability does not constitute a necessary condition for legitimate statistical explanations. Taking them together with the vitamin C example, we must conclude—provisionally, at least—that a high probability of the explanandum relative to the explanans is neither necessary nor sufficient for correct statistical explanations, even if all of Hempel's other conditions are fulfilled. Much more remains to be said about the high-probability requirement, for it raises a host of fundamental philosophical problems, but I shall postpone further discussion of it until chapter 4.

Given the problematic status of the high-probability requirement, it was natural to attempt to construct an alternative treatment of statistical explanation that rests upon different principles. As I argued in (Salmon, 1965), statistical relevance, rather than high probability, seems to be the key explanatory relationship. This starting point leads to a conception of scientific explanation that differs fundamentally and radically from Hempel's I-S account. In the first place, if we are to make use of statistical relevance relations, our explanations will have to make reference to at least two probabilities, for statistical relevance involves a difference between two probabilities. More precisely, a factor C is statistically relevant to the occurrence of B under circumstances A if and only if

$$P(B|A.C) \neq P(B|A) \tag{1}$$

or

$$P(B|A.C) \neq P(B|A.\bar{C}). \tag{2}$$

Conditions (1) and (2) are equivalent to one another, provided that C occurs with a nonvanishing probability within A; since we shall not be concerned with the relevance of factors whose probabilities are zero, we may use either (1) or (2) as our definition of statistical relevance. We say that C is positively relevant to B if the probability of B is greater in the presence of C; it is negatively relevant if the probability of B is smaller in the presence of C. For instance, heavy cigarette smoking is positively relevant to the occurrence of lung cancer, at some later time, in a thirty-year-old Australian male; it is negatively relevant to survival to the age of seventy for such a person.

In order to construct a satisfactory statistical explanation, it seems to me, we need a *prior probability* of the occurrence to be explained, as well as one or more *posterior probabilities*. A crucial feature of the explanation will be the comparison between the prior and posterior probabilities. In Hempel's case of the streptococcus infection, for instance, we might begin with the probability, in the entire class of people with streptococcus infections, of a quick recovery. We realize, however, that the administration of penicillin is statistically relevant to quick recovery, so we compare the probability of quick recovery among those who have received penicillin with the probability of quick recovery among those who have not received penicillin. Hempel warns, however, that there is another relevant factor, namely, the existence of the penicillin-resistant strain of bacteria. We must, therefore, take that factor into account as well. Our original reference class has been divided into four parts: (1) infection by non-penicillin-resistant bacteria, penicillin given; (2) infection by non-penicillin-resistant bacteria, no penicillin given; (3) infection by penicillin-resistant bacteria, penicillin given; (4) infection by penicillin-resistant bacteria, no penicillin given. Since the administration of penicillin is irrelevant to quick recovery in case of penicillin-resistant infections, the subclasses (3) and (4) of the original reference class should be merged to yield (3') infection by penicillin-resistant bacteria. If John Jones is a member of (1), we have an explanation of his quick recovery, according to the S-R approach, not because the probability is high, but, rather, because it differs significantly from the probability in the original reference class. We shall see later what must be done if John Jones happens to fall into class (3').

By way of contrast, Hempel's earlier high-probability requirement demands only that the posterior probability be sufficiently large—whatever that might mean—but makes no reference at all to any prior probability. According to Hempel's abstract model, we ask, "Why is individual x a member of B?" The answer consists of an inductive argument having the following form:

$$P(B|A) = r$$
$$x \text{ is an } A$$
$$\rule{3cm}{0.4pt} \quad [r]$$
$$x \text{ is a } B$$

As we have seen, even if the first premise is a statistical law, r is high, the premises are true, and the requirement of maximal specificity has been fulfilled, our 'explanation' may be patently inadequate, due to failure of relevancy.

In (Salmon, 1970, pp. 220–221), I advocated what came to be called the statistical-relevance or S-R model of scientific explanation. At that time, I thought that anything that satisfied the conditions that define that model would qualify as a legitimate scientific explanation. I no longer hold that view. It now seems to me that the statistical relationships specified in the S-R model constitute the *statistical basis* for a bona fide scientific explanation, but that this basis must be supplemented by certain *causal factors* in order to constitute a satisfactory scientific explanation. In chapters 5–9 I shall discuss the causal aspects of explanation. In this chapter, however, I shall confine attention to the statistical basis, as articulated in terms of the S-R model. Indeed, from here on I shall speak, not of the S-R model, but, rather, of the *S-R basis*.[10]

Adopting the S-R approach, we begin with an explanatory question in a form somewhat different from that given by Hempel. Instead of asking, for instance, "Why did x get well within a fortnight?" we ask, "Why did this person with a cold get well within a fortnight?" Instead of asking, "Why is x a B?" we ask, "Why is x, which is an A, also a B?" The answer—at least for preliminary purposes—is that x is also a C, where C

---

[10] I am extremely sympathetic to the thesis, expounded in (Humphreys, 1983), that probabilities—including those appearing in the S-R basis—are important tools in the construction of scientific explanations, but that they do not constitute any part of a scientific explanation per se. This thesis allows him to relax considerably the kinds of maximal specificity or homogeneity requirements that must be satisfied by statistical or probabilistic explanations. A factor that is statistically relevant may be causally irrelevant because, for example, it does not convert any contributing causes to counteracting causes or vice versa. This kind of relaxation is attractive in a theory of scientific explanation, for factors having small statistical relevance often seem otiose. Humphreys' approach does not show, however, that such relevance relations can be omitted from the S-R basis; on the contrary, the S-R basis must include such factors in order that we may ascertain whether they can be omitted from the causal explanation or not. I shall return to Humphreys' concept of aleatory explanation in chapter 9.

is *relevant* to B within A. Thus we have a prior probability $P(B|A)$—in this case, the probability that a person with a cold (A) gets well within a fortnight (B). Then we let C stand for the taking of vitamin C. We are interested in the posterior probability $P(B|A.C)$ that a person with a cold who takes vitamin C recovers within a fortnight. If the prior and posterior probabilities are equal to one another, the taking of vitamin C can play no role in explaining why this person recovered from the cold within the specified period of time. If the posterior probability is not equal to the prior probability, then C may, under certain circumstances, furnish part or all of the desired explanation. A large part of the purpose of the present book is to investigate the way in which considerations that are statistically relevant to a given occurrence have or lack explanatory import.

We cannot, of course, expect that every request for a scientific explanation will be phrased in canonical form. Someone might ask, for example, "Why did Mary Jones get well in no more than a fortnight's time?" It might be clear from the context that she was suffering from a cold, so that the question could be reformulated as, "Why did this person who was suffering from a cold get well within a fortnight?" In some cases, it might be necessary to seek additional clarification from the person requesting the explanation, but presumably it will be possible to discover what explanation is being called for. This point about the form of the explanation-seeking question has fundamental importance. We can easily imagine circumstances in which an altogether different explanation is sought by means of the same initial question. Perhaps Mary had exhibited symptoms strongly suggesting that she had mononucleosis; in this case, the fact that it was only an ordinary cold might constitute the explanation of her quick recovery. A given why-question, construed in one way, may elicit an explanation, while otherwise construed, it asks for an explanation that cannot be given. "Why did the Mayor contract paresis?" might mean, "Why did this adult human develop paresis?" or, "Why did this syphilitic develop paresis?" On the first construal, the question has a suitable answer, which we have already discussed. On the second construal, it has no answer— at any rate, we cannot give an answer—for we do not know of any fact in addition to syphilis that is relevant to the occurrence of paresis. Some philosophers have argued, because of these considerations, that scientific explanation has an unavoidably pragmatic aspect (e.g., van Fraassen, 1977, 1980). If this means simply that there are cases in which people ask for explanations in unclear or ambiguous terms, so that we cannot tell what explanation is being requested without further clarification, then so be it. No one would deny that we cannot be expected to supply explanations unless we know what it is we are being asked to explain. To this extent, scientific explanation surely has pragmatic or contextual components. Dealing with these considerations is, I believe, tantamount to choosing a suitable reference class with respect to which the prior probabilities are to be taken and specifying an appropriate sample space for purposes of a particular explanation. More will be said about these two items in the next section. In chapter 4—in an extended discussion of van Fraassen's theory—we shall return to this issue of pragmatic aspects of explanation, and we shall consider the question of whether there are any others.

The Statistical-Relevance Approach

Let us now turn to the task of giving a detailed elaboration of the S-R basis. For purposes of initial presentation, let us construe the terms A, B, C, . . . (with or without subscripts) as referring to classes, and let us construe our probabilities in some sense as relative frequencies. This *does not mean* that the statistical-relevance approach is tied in any crucial way to a frequency theory of probability. I am simply adopting the heuristic device of picking examples involving frequencies because they are easily grasped. Those who prefer propensities, for example, can easily make the appropriate terminological adjustments, by speaking of chance setups and outcomes of trials where I refer to reference classes and attributes. With this understanding in mind, let us consider the steps involved in constructing an S-R basis for a scientific explanation:

1. We begin by selecting an appropriate reference class A with respect to which the prior probabilities $P(B_i|A)$ of the B_is are to be taken.
2. We impose an *explanandum-partition* upon the initial reference class A in terms of an exclusive and exhaustive set of attributes B_1, . . . , B_m; this defines a sample space for purposes of the explanation under consideration. (This partition was not required in earlier presentations of the S-R model.)
3. Invoking a set of statistically relevant factors C_1, . . . , C_s, we partition the initial reference class A into a set of mutually exclusive and exhaustive cells $A.C_1$, . . . , $A.C_s$. The properties C_1, . . . , C_s furnish the *explanans-partition*.
4. We ascertain the associated probability relations:
    prior probabilities

    $P(B_i|A) = p_i$
      for all i $(1 \leq i \leq m)$

    posterior probabilities

    $P(B_i|A.C_j) = p_{ij}$
      for all i and j $(1 \leq i \leq m)$ and $(1 \leq j \leq s)$

5. We require that each of the cells $A.C_j$ be homogeneous with respect to the explanandum-partition $\{B_i\}$; that is, none of the cells in the partition can be further subdivided in any manner relevant to the occurrence of any B_i. (This requirement is somewhat analogous to Hempel's requirement of maximal specificity, but as we shall see, it is a much stronger condition.)

6. We ascertain the relative sizes of the cells in our explanans-partition in terms of the following marginal probabilities:

$$P(C_j|A) = q_j$$

(These probabilities were not included in earlier versions of the S-R model; the reasons for requiring them now will be discussed later in this chapter.)

7. We require that the explanans-partition be a maximal homogeneous partition, that is—with an important exception to be noted later—for $i \neq k$ we require that $p_{ji} \neq p_{jk}$. (This requirement assures us that our partition in terms of $C_1, \ldots, C_m$ does not introduce any irrelevant subdivision in the initial reference class A.)

8. We determine which cell $A.C_j$ contains the individual x whose possession of the attribute B_i was to be explained. The probability of B_i within the cell is given in the list under 4.

Consider in a rather rough and informal manner the way in which the foregoing pattern of explanation might be applied in a concrete situation; an example of this sort was offered by James Greeno (1971a, pp. 89–90). Suppose that Albert has committed a delinquent act—say, stealing a car, a major crime—and we ask for an explanation of that fact. We ascertain from the context that he is an American teen-ager, and so we ask, "Why did this American teen-ager commit a serious delinquent act?" The prior probabilities, which we take as our point of departure, so to speak, are simply the probabilities of the various degrees of juvenile delinquency (B_i) among American teen-agers (A)—that is, $P(B_i|A)$. We will need a suitable explanandum-partition; Greeno suggests B_1 = no criminal convictions, B_2 = conviction for minor infractions only, B_3 = conviction for a major offense. Our sociological theories tell us that such factors as sex, religious background, marital status of parents, type of residential community, socioeconomic status, and several others are relevant to delinquent behavior. We therefore take the initial reference class of American teen-agers and divide it into males and females; Jewish, Protestant, Roman Catholic, no religion; parents married, parents divorced, parents never married; urban, suburban, rural place of residence; upper, middle, lower class; and so forth. Taking such considerations into account, we arrive at a large number *s* of cells in our partition. We assign probabilities of the various degrees of delinquent behavior to each of the cells in accordance with 4, and we ascertain the probability of a randomly selected American teen-ager belonging to each of the cells in accordance with 6. We find the cell to which Albert belongs—for example, male, from a Protestant background, parents divorced, living in a suburban area, belonging to the middle class. If we have taken into account all of the relevant factors, and if we have correctly ascertained the probabilities associated with the various cells of our partitions, then we have an S-R basis for the explanation of Albert's delinquency that conforms to the foregoing schema. If it should turn out (contrary to what I believe actually to be the case) that the probabilities of the various types of delinquency are the same for males and for females, then we would not use sex in partitioning our original reference class. By condition 5 we must employ *every* relevant factor; by condition 7 we must employ *only* relevant factors. In many concrete situations, including the present examples, we know that we have not found all relevant considerations; however, as Noretta Koertge rightly emphasized (1975), that is an ideal for which we may aim. Our philosophical analysis is designed to capture the notion of a fully satisfactory explanation.

Nothing has been said, so far, concerning the rationale for conditions 2 and 6, which are here added to the S-R basis for the first time. We must see why these requirements are needed. Condition 2 is quite straightforward; it amounts to a requirement that the sample space for the problem at hand be specified. As we shall see when we discuss Greeno's information-theoretic approach in chapter 4, both the explanans-partition and the explanandum-partition are needed to measure the information transmitted in any explanatory scheme. This is a useful measure of the explanatory value of a theory. In addition, as we shall see when we discuss van Fraassen's treatment of why-questions in chapter 4, his contrast class, which is the same as our explanandum-partition, is needed in some cases to specify precisely what explanation is being sought. In dealing with the question "Why did Albert steal a car?" we used Greeno's suggested explanandum-partition. If, however, we had used different partitions (contrast classes), other explanations might have been called forth. Suppose that the contrast class included: Albert steals a car, Bill steals a car, Charlie steals a car, and so forth. Then the answer might have involved no sociology whatever; the explanation might have been that, among the members of his gang, Albert is most adept at hot-wiring. Suppose, instead, that the contrast class had included: Albert steals a car, Albert steals a diamond ring, Albert steals a bottle of whiskey, and so forth. In that case, the answer might have been that he wanted to go joyriding.

The need for the marginal probabilities mentioned in 6 arises in the following way. In many cases, such as the foregoing delinquency example, the terms C_j that furnish the explanans-partition of the initial reference class are conjunctive. A given cell is determined by several distinct factors: sex *and* religious background *and* marital status of parents *and* type of residential community *and* socioeconomic status *and* . . . which may be designated D_k, E_n, F_r, ... These factors will be the probabilistic contributing causes and counteracting causes that tend, respectively, to produce or prevent delinquency. In attempting to understand the phenomenon in question, it is important to know how each factor is relevant—whether positively or negatively, and how strongly—both in the population at large and in various subgroups of the population. Consider, for example, the matter of sex. It may be that within the entire class of American teenagers (A) the probability of serious delinquency (B_3) is greater among males than it is among females. If so, we would want to know by how much the probability among males exceeds the probability among females and by how much it exceeds the probability in the entire population. We also want to know whether being male is always positively relevant to serious delinquency, or whether in combination with certain other factors, it may be negatively relevant or irrelevant. Given two groups of teen-agers—one consisting entirely of boys and the other entirely of girls, but alike with respect to all of the other factors—we want to know how the probabilities associated with delinquency in each of the two groups are related to one another. It might be that in each case of two cells in the explanandum-partition that differ from one another only on the basis of gender, the probability of serious delinquency in the male group is greater than it is in the female group. It might turn out, however, that sometimes the two probabilities are equal, or that in some cases the probability is higher in the female group than it is in the corresponding male group. Relationships of all of these kinds are logically possible.

It is a rather obvious fact that each of two circumstances can individually be positively relevant to a given outcome, but their conjunction can be negatively relevant or irrelevant. Each of two drugs can be positively relevant to good health, but taken together, the combination may be detrimental—for example, certain antidepressive medications taken in conjunction with various remedies for the common cold can greatly increase the chance of dangerously high blood pressure (Goodwin and Guze, 1979). A factor that is a contributing cause in some circumstances can be a counteracting cause in other cases. Problems of this sort have been discussed, sometimes under the heading of "Simpson's paradox," by Nancy Cartwright (1983, essay 1) and Bas van Fraassen (1980, pp. 108, 148–151). In (Salmon, 1975c), I have spelled out in detail the complexities that arise in connection with statistical relevance relations. The moral is that we need to know not only how the various factors D_k, E_n, F_r, . . . , are relevant to the outcome, B_i, but how the relevance of each of them is affected by the presence or absence of the others. Thus, for instance, it is possible that being female might in general be negatively relevant to delinquency, but it might be positively relevant among the very poor.

Even if all of the prior probabilities $P(B_i|A)$ and all of the posterior probabilities $P(B_i|A.C_j)$ furnished under condition 4 are known, it is not possible to deduce the conditional probabilities of the B_i's with respect to the individual conjuncts that make up the C_j's or with respect to combinations of them. Without these conditional probabilities, we will not be in a position to ascertain all of the statistical relevance relations that are required. We therefore need to build in a way to extract that information. This is the function of the marginal probabilities $P(C_j|A)$ required by condition 6. If these are known, such conditional probabilities as $P(B_i|A.D_k)$, $P(B_i|A.E_n)$, and $P(B_i|A.D_k.E_n)$ can be derived.[11] When 2 and 6 are added to the earlier characterization of the S-R model (Salmon et al., 1971), then, I believe, we have gone as far as possible in characterizing scientific explanations at the level of statistical relevance relations.

[11] Suppose, for example, that we wish to compute $P(B_i|A.D_k)$, where $D_k = C_{j_1} \vee \ldots \vee C_{j_q}$, the cells C_{j_r} being mutually exclusive. This can be done as follows. We are given $P(C_j|A)$ and $P(B_i|A.C_j)$. By the multiplication theorem,

$$P(D_k.B_i|A) = P(D_k|A) \times P(B_i|A.D_k)$$

Assuming $P(D_k|A) \neq 0$, we have,

$$P(B_i|A.D_k) = P(D_k.B_i|A)/P(D_k|A) \qquad (*)$$

By the addition theorem

$$P(D_k \mid A) = \sum_{r=1}^{q} P(C_{j_r} \mid A)$$

$$P(D_k.B_i \mid A) = \sum_{r=1}^{q} P(C_{j_r}.B_i \mid A)$$

By the multiplication theorem

$$P(D_k.B_i \mid A) = \sum_{r=1}^{q} P(C_{j_r} \mid A) \times P(B_i \mid A.C_{j_r})$$

Substitution in (*) yields the desired relation:

$$P(B_i \mid A.D_k) = \frac{\sum_{r=1}^{q} P(C_{j_r} \mid A) \times P(B_i \mid A.C_{j_r})}{\sum_{r=1}^{q} P(C_{j_r} \mid A)}$$

Several features of the new version of the S-R basis deserve explicit mention. It should be noted, in the first place, that conditions 2 and 3 demand that the entire initial reference class A be partitioned, while conditions 4 and 6 require that *all* of the associated probability values be given. This is one of several respects in which it differs from Hempel's I-S model. Hempel requires only that the individual mentioned in the explanandum be placed within an appropriate class, satisfying his requirement of maximal specificity, but he does not ask for information about any class in either the explanandum-partition or the explanans-partition to which that individual does not belong. Thus he might go along in requiring that Bill Smith be referred to the class of American male teen-agers coming from a Protestant background, whose parents are divorced, and who is a middle-class suburban dweller, and in asking us to furnish the probability of his degree of delinquency within that class. But why, it may surely be asked, should we be concerned with the probability of delinquency in a lower-class, urban-American, female teen-ager from a Roman Catholic background whose parents are still married? What bearing do such facts have on Bill Smith's delinquency? The answer, I think, involves serious issues concerning scientific generality. If we ask why this American teen-ager becomes a delinquent, then, it seems to me, we are concerned with *all* of the factors that are relevant to the occurrence of delinquency, and with the ways in which these factors are relevant to that phenomenon (cf. Koertge, 1975). To have a satisfactory scientific answer to the question of why this A is a B_i—to achieve full scientific understanding—we need to know the factors that are relevant to the occurrence of the various B_i s for *any* randomly chosen or otherwise unspecified member of A. It was mainly to make good on this desideratum that requirement 6 was added. Moreover, as Greeno and I argued in *Statistical Explanation and Statistical Relevance*, a good measure of the value of an S-R basis is the gain in information furnished by the complete partitions and the associated probabilities. This measure cannot be applied to the individual cells one at a time.

A fundamental philosophical difference between our S-R basis and Hempel's I-S model lies in the interpretation of the concept of homogeneity that appears in condition 5. Hempel's requirement of maximal specificity, which is designed to achieve a certain kind of homogeneity in the reference classes employed in I-S explanations, is *epistemically relativized*. This means, in effect, that we must not *know* of any way to make a relevant partition, but it certainly does not demand that no possibility of a relevant partition can exist unbeknown to us. As I view the S-R basis, in contrast, condition 5 demands that the cells of our explanans-partition be *objectively* homogeneous; for this model, homogeneity is not epistemically relativized. Since this issue of epistemic relativization versus objective homogeneity is discussed at length in chapter 3, it is sufficient for now merely to call attention to this complex problem.[12]

Condition 7 has been the source of considerable criticism. One such objection rests on the fact that the initial reference class A, to which the S-R basis is referred, may not be maximal. Regarding Kyburg's hexed salt example, mentioned previously, it has been pointed out that the class of samples of table salt is not a maximal homogeneous class with respect to solubility, for there are many other chemical substances that have the same probability—namely, unity—of dissolving when placed in water. Baking soda, potassium chloride, various sugars, and many other compounds have this property. Consequently, if we take the maximality condition seriously, it has been argued, we should not ask, "Why does this sample of table salt dissolve in water?" but, rather, "Why does this sample of matter in the solid state dissolve when placed in water?" And indeed, one can argue, as Koertge has done persuasively (1975), that to follow such a policy often leads to significant scientific progress. Without denying her important point, I would nevertheless suggest, for purposes of elaborating the formal schema, that we take the initial reference class A as given by the explanation-seeking why-question, and look for relevant partitions within it. A significantly different explanation, which often undeniably represents scientific progress, may result if a different why-question, embodying a broader initial reference class, is posed. If the original question is not presented in a form that unambiguously determines a reference class A, we can reasonably discuss the advantages of choosing a wider or a narrower class in the case at hand.

Another difficulty with condition 7 arises if 'accidentally'—so to speak— two different cells in the partition, $A.C_i$ and $A.C_j$, happen to have equal associated probabilities p_{ki} and p_{kj} for all cells B_k in the explanandum-

---

[12] Cartwright (1983, p. 27) asserts that on Hempel's account, "what counts as a good explanation is an objective, person-independent matter," and she applauds him for holding that view. I find it difficult to reconcile her characterization with Hempel's repeated emphatic assertion (prior to 1977) of the doctrine of essential epistemic relativization of inductive-statistical explanation. In addition, she complains that my way of dealing with problems concerning the proper formulation of the explanation-seeking why-question—that is, problems concerning the choice of an appropriate initial reference class—"makes explanation a subjective matter" (ibid., p. 29). "What explains what," she continues, "depends on the laws and facts true in our world, and cannot be adjusted by shifting our interest or our focus" (ibid.). This criticism seems to me to be mistaken. Clarification of the question is often required to determine what it is that is to be explained, and this may have pragmatic dimensions. However, once the explanandum has been unambiguously specified, on my view, the identification of the appropriate explanans is fully objective. I am in complete agreement with Cartwright concerning the desirability of such objectivity; moreover, my extensive concern with objective homogeneity is based directly upon the desire to eliminate from the theory of statistical explanation such subjective features as epistemic relativization.

partition. Such a circumstance might arise if the cells are determined conjunctively by a number of relevant factors, and if the differences between the two cells cancel one another out. It might happen, for example, that the probabilities of the various degrees of delinquency—major offense, minor offense, no offense—for an upper-class, urban, Jewish girl would be equal to those for a middle-class, rural, Protestant boy. In this case, we might want to relax condition 7, allowing $A.C_i$ and $A.C_j$ to stand as separate cells, provided they differ with respect to at least two of the terms in the conjunction, so that we are faced with a fortuitous canceling of relevant factors. If, however, $A.C_i$ and $A.C_j$ differed with respect to only one conjunct, they would have to be merged into a single cell. Such would be the case if, for example, among upper-class, urban-dwelling, American teen-agers whose religious background is atheistic and whose parents are divorced, the probability of delinquent behavior were the same for boys as for girls. Indeed, we have already encountered this situation in connection with Hempel's example of the streptococcus infection. If the infection is of the penicillin-resistant variety, the probability of recovery in a given period of time is the same whether penicillin is administered or not. In such cases, we want to say, there is no relevant difference between the two classes—not that relevant factors were canceling one another out. I bring this problem up for consideration at this point, but I shall not make a consequent modification in the formal characterization of the S-R basis, for I believe that problems of this sort are best handled in the light of causal relevance relations. Indeed, it seems advisable to postpone detailed consideration of the whole matter of regarding the cells $A.C_j$ as being determined conjunctively until causation has been explicitly introduced into the discussion. As we shall see in chapter 9 (Humphreys, 1981, 1983) and (Rogers, 1981) provide useful suggestions for handling just this issue.

Perhaps the most serious objection to the S-R model of scientific explanation—as it was originally presented—is based upon the principle that *mere* statistical correlations explain nothing. A rapidly falling barometric reading is a sign of an imminent storm, and it is *highly correlated* with the onset of storms, but it certainly does not *explain* the occurrence of a storm. The S-R approach does, however, have a way of dealing with examples of this sort. A factor C, which is relevant to the occurrence of B in the presence of A, may be screened off in the presence of some additional factor D; the screening-off relation is defined by equations (3) and (4), which follow. To illustrate, given a series of days (A) in some particular locale, the probability of a storm occurring (B) is in general quite different from the probability of a storm if there has been a recent sharp drop in the barometric reading (C). Thus C is statistically relevant to B within A. If, however, we take into consideration the further fact that

relations—with the screening-off relation playing a crucial role. As I shall explain in chapter 6, I no longer believe this is possible. A large part of the material in the present book is devoted to an attempt to analyze the nature of the causal relations that enter into scientific explanations, and the manner in which they function in explanatory contexts. After characterizing the S-R model, I wrote:

> One might ask on what grounds we can claim to have characterized explanation. The answer is this. When an explanation (as herein explicated) has been provided, we know exactly how to regard any A with respect to the property B. We know which ones to bet on, which to bet against, and at what odds. We know precisely what degree of expectation is rational. We know how to face uncertainty about an A's being a B in the most reasonable, practical, and efficient way. We know every factor that is relevant to an A having the property B. We know exactly what weight should have been attached to the prediction that this A will be a B. We know all of the regularities (universal and statistical) that are relevant to our original question. What more could one ask of an explanation? (Salmon et al., 1971, p. 78)

The answer, of course, is that we need to know something about the causal relationships as well.

In acknowledging this egregious shortcoming of the S-R model of scientific explanation, I am not abandoning it completely. The attempt, rather, is to supplement it in suitable ways. While recognizing its incompleteness, I still think it constitutes a sound basis upon which to erect a more adequate account. And at a fundamental level, I still think it provides important insights into the nature of scientific explanation.

In the introduction to *Statistical Explanation and Statistical Relevance*, I offered the following succinct comparison between Hempel's I-S model and the S-R model:

I-S model: an explanation is an *argument* that renders the explanandum *highly probable*.

S-R model: an explanation is an *assembly of facts statistically relevant to the explanandum, regardless of the degree of probability that results*.

It was Richard Jeffrey (1969) who first explicitly formulated the thesis that (at least some) statistical explanations are not arguments; it is beautifully expressed in his brief paper, "Statistical Explanation vs. Statistical Inference," which was reprinted in *Statistical Explanation and Statistical Relevance*. In (Salmon, 1965, pp. 145–146), I had urged that positive relevance rather than high probability is the desideratum in statistical explanation.

In (Salmon, 1970), I expressed the view, which many philosophers found weird and counter-intuitive (e.g., L. J. Cohen, 1975), that statistical explanations may even embody *negative* relevance—that is, an explanation of an event may, in some cases, show that the event to be explained is less probable than we had initially realized. I still do not regard that thesis as absurd. In an illuminating discussion of the explanatory force of positively and negatively relevant factors, Paul Humphreys (1981) has introduced some felicitous terminology for dealing with such cases, and he has pointed to an important constraint. Consider a simple example. Smith is stricken with a heart attack, and the doctor says, "*Despite* the fact that Smith exercised regularly and had given up smoking several years ago, he contracted heart disease *because* he was seriously overweight." The "because" clause mentions those factors that are positively relevant and the "despite" clause cites those that are negatively relevant. Humphreys refers to them as *contributing causes* and *counteracting causes*, respectively. When we discuss causal explanation in later chapters, we will want to say that a complete explanation of an event must make mention of the causal factors that tend to prevent its occurrence as well as those that tend to bring it about. Thus it is *not* inappropriate for the S-R basis to include factors that are negatively relevant to the explanandum-event. As Humphreys points out, however, we would hardly consider as appropriate a putative explanation that had only negative items in the "despite" clause and no positive items in the "because" category. "Despite the fact that Jones never smoked, exercised regularly, was not overweight, and did not have elevated levels of triglycerides and cholesterol, he died of a heart attack," would hardly be considered an acceptable *explanation* of his fatal illness.

Before concluding this chapter on models of statistical explanation, we should take a brief look at the deductive-homological-probabilistic (D-N-P) model of scientific explanation offered by Peter Railton (1978). By employing well-established statistical laws, such as that covering the spontaneous radioactive decay of unstable nuclei, it is possible to deduce the fact that a decay-event for a particular isotope has a certain probability of occurring within a given time interval. For an atom of carbon 14 (which is used in radiocarbon dating in archaeology, for example), the probability of a decay in 5,730 years is 1/2. The explanation of *the probability of the decay-event* conforms to Hempel's deductive-nomological pattern. Such an explanation does not, however, explain the actual occurrence of a decay, for, given the probability of such an event—however high or low—the event in question may not even occur. Thus the explanation does not qualify as an argument to the effect that the event-to-be-explained was to be expected with deductive certainty, given the explanans. Railton is, of course, clearly aware of the fact. He goes on to point out, nevertheless, that if we simply attach an "addendum" to the deductive argument stating that the event-to-be-explained did, in fact, occur in the case at hand, we can claim to have a probabilistic *account*—which is not a deductive or inductive argument—of the occurrence of the event. In this respect, Railton is in rather close agreement with Jeffrey (1969) that some explanations are not arguments. He also agrees with Jeffrey in emphasizing the importance of exhibiting the physical mechanisms that lead up to the probabilistic occurrence that is to be explained. Railton's theory—like that of Jeffrey—has some deep affinities to the S-R model. In including a reference to physical mechanisms as an essential part of his D-N-P model, however, Railton goes beyond the view that statistical relevance relations, in and of themselves, have explanatory import. His theory of scientific explanation can be more appropriately characterized as causal or mechanistic. It is closely related to the two-tiered causal-statistical account that I am attempting to elaborate as an improvement upon the S-R model.

Although, with Kyburg's help, I have offered what seem to be damaging counterexamples to the D-N model—for instance, the one about the man who explains his own avoidance of pregnancy on the basis of his regular consumption of his wife's birth control pills (Salmon et al., 1971, p. 34)—the major emphasis has been upon statistical explanation, and that continues to be the case in what follows. Aside from the fact that contemporary science obviously provides many statistical explanations of many types of phenomena, and that any philosophical theory of statistical explanation has only lately come forth, there is a further reason for focusing upon statistical explanation. As I maintained in chapter 1, we can identify three distinct approaches to scientific explanation that do not seem to differ from one another in any important way as long as we confine our attention to contexts in which all of the explanatory laws are universal generalizations. I shall argue in chapter 4, however, that these three general conceptions of scientific explanation can be seen to differ radically from one another when we move on to situations in which statistical explanations are in principle the best we can achieve. Close consideration of statistical explanations, with sufficient attention to their causal ingredients, provides important insight into the underlying philosophical questions relating to our scientific understanding of the world.

3 | Objective Homogeneity

IN THE preceding chapter, we discussed two quite different models of statistical explanation in some detail. One of these—the I-S model—fits clearly within the epistemic conception of scientific explanation. The other—the S-R model—was designed to exemplify the ontic conception. The fact that two divergent basic conceptions are represented suggests that fundamental philosophical issues are involved. This chapter will focus upon one of these issues, namely, the problem of epistemic relativization. In order to deal with this problem, we shall find it necessary to provide a detailed analysis of the concept of homogeneity of reference classes—a concept that turns out to be tantamount to the notion of physical randomness.

EPISTEMIC RELATIVIZATION

As we have seen, Hempel regarded the ambiguity of inductive-statistical explanation as a cardinal difficulty to be overcome in constructing a satisfactory theory of statistical explanation.[1] The requirement of maximal specificity was introduced expressly to do this job. Since this requirement, which must be satisfied by *every* I-S explanation, is formulated in terms of particular knowledge situations, Hempel was led to enunciate the thesis of *epistemic relativity of statistical explanation*. This thesis asserts "that *the concept of statistical explanation for particular events is essentially relative to a given knowledge situation*" (1965, pp. 402–403, italics in original). The entire tenor of his discussion, in addition to the occurrence of the word "essential" in his formulation of the thesis, seems clearly to imply that, on Hempel's view (1965, 1968), there is not and cannot be any such thing as an I-S explanation *simpliciter*, but only an I-S explanation *relative to a given knowledge situation*.

Hempel's I-S pattern of explanation of particular events requires incorporation of at least one statistical generalization in the explanans. For

[1] In an article provocatively entitled "A Deductive-Nomological Model of Probabilistic Explanation," Peter Railton (1978) offers an account of probabilistic explanation explicitly designed to escape the problems of maximal specificity and epistemic relativization. Whether his attempt is successful or not, it does not enable Hempel's *inferential* theory to avoid these difficulties, for the price Railton pays is to give up the thesis that explanations are arguments (pp. 209, 217).

reasons already discussed in chapter 2, it is obvious that the reference class mentioned in the statistical generalization must satisfy some sort of condition at least closely akin to the requirement of maximal specificity. The requirement, stated very roughly, demands that no way be known of making a subdivision of the reference class that is statistically relevant to the occurrence of the explanandum-event. Since what is known is obviously a function of a body of knowledge, the question of whether a given reference class satisfies the requirement of maximal specificity can be meaningful only with respect to a specific knowledge situation.

Let us say that a reference class is *epistemically homogeneous* with respect to a given attribute—relative to a given knowledge situation—if no way is known to make a relevant partition of it. Let us say that a reference class is *objectively homogeneous* with respect to a given attribute if there is in fact no way of effecting a relevant partition.[2] Objective homogeneity is obviously not relativized to any particular knowledge situation. Objective homogeneity can be illustrated by trivial examples. If it is true that all objects composed of pure silver are electrical conductors, then the class of objects composed of pure silver is objectively homogeneous with respect to the attribute of being an electrical conductor. If every member of a given class possesses a given attribute, then according to a logical truism, every member of any subclass of that given class will possess the attribute in question. It is possible, of course, that we are mistaken in our belief that every object made of pure silver is an electrical conductor. The fact that we *might be* mistaken in this conviction does not mean that the class of objects composed of pure silver *is* merely epistemically homogeneous and not objectively homogeneous. The only conclusion that can be drawn from this possibility is that we might be mistaken in claiming that the class in question is objectively homogeneous. Whether silver is or is not an electrical conductor is a question of objective fact that is entirely independent of anyone's state of knowledge—indeed, it does not even depend upon the existence of intelligent beings.

It is of crucial importance to be clear on this point. The statement that all objects composed of pure silver are electrical conductors is a universal generalization, not a statistical generalization. It is the type of universal law-statement that could appear in the explanans of a D-N explanation. In full awareness of the fact that scientific generalizations are never established with absolute certainty, and that we might be mistaken in any instance in which we assert a universal generalization as expressing a law of nature,

[2] Homogeneity, whether epistemic or objective, is obviously relative to a specified attribute. This point is taken as understood throughout the present discussion, whether it happens to be mentioned explicitly or not.

Hempel has explicitly and repeatedly emphasized his view that D-N explanations are *not* essentially relativized to knowledge situations. When we offer a D-N explanation, we depend, of course, on what we know (or think we know), but that fact does not render a D-N explanation "essentially relative to a given knowledge situation" (1965, p. 403). It merely means that we may be in error when we believe we have a D-N explanation of some fact or other. It is not conceptually incoherent to claim that we have a D-N explanation *simpliciter*, even though that claim may be in error.

When considering the kinds of statistical generalizations employed in I-S explanations, Hempel correctly observes that an *epistemically* homogeneous reference class may be objectively inhomogeneous (1965, p. 402). Thus, it would seem, we might sometimes claim that a particular reference class mentioned in a statistical generalization is objectively homogeneous when it is in fact inhomogeneous. That is to say, we might believe that there is no relevant partition that can be made in a reference class when there actually exists a relevant partition of which we are unaware. This sort of factual error is completely on a par, I should think, with the factual error we would make in asserting a universal generalization that is false. Since the ever-present possibility of errors of the latter kind does not render a D-N explanation essentially relative to a given knowledge situation, we must consider carefully whether the ever-present possibility of the former type of error does provide adequate ground for maintaining that I-S explanations are *essentially* relativized to a given knowledge situation.

Perhaps the best way to pose the question is this. Suppose we had a set of statements, including one statistical generalization, that fulfill Hempel's requirements for an I-S explanation (in a given knowledge situation). Suppose, further, that someone were to assert that the reference class mentioned in the statistical generalization is not merely epistemically homogeneous, but also objectively homogeneous. If this latter assertion could be justified, then I should think we could claim to have an I-S explanation *simpliciter*—one that is not essentially relativized to any knowledge situation. Hempel maintains, however, that *all* I-S explanations are essentially relativized to knowledge situations, from which it follows that we can never, in his opinion, be justified in asserting the objective homogeneity of any reference class mentioned in a (nonuniversal) statistical generalization.

There are two grounds that might be invoked for denying that we could ever be justified in asserting that a given reference class is objectively homogeneous:

(1) It might be argued that the claim is always false, and that the only cases of objectively homogeneous reference classes are those that are

trivially homogeneous because they occur in universal (nonstatistical) generalizations. Consider, for instance, the paresis example, which was introduced in the preceding chapter, as a case of explanation that does not satisfy the high-probability requirement. An obvious response to that claim would have been to maintain that although the presence of latent untreated syphilis provides some sort of partial explanation of paresis, we simply do not know enough at present to provide anything like an adequate explanation of paresis. As medical knowledge increases, it might have been argued, we expect to find some additional factor or set of factors F such that those victims of latent untreated syphilis in which F is present will very probably develop paresis, while those in which F is absent will very probably not develop paresis. With this hypothetical improvement in our knowledge situation, we could then claim to have an I-S explanation that satisfies all of Hempel's conditions, including the high-probability requirement. As long as we lack knowledge of F, we can supply only a partial explanation, but when we discover F, we can offer a satisfactory I-S explanation of an instance of paresis.

We must not, however, lose sight of the fact that the resulting I-S explanation is relativized to a knowledge situation—that is, it is an adequate I-S explanation with respect to a given body of knowledge K, but it cannot be said to be an adequate explanation simpliciter. According to Hempel's thesis of essential epistemic relativization, there are no adequate I-S explanations simpliciter. But if the reference class appearing in the statistical law—the class of people with latent untreated syphilis (S) who also possess the factor F—were objectively homogeneous, then we would have an I-S explanation simpliciter. Consequently, the reference class $S.F$ must be objectively inhomogeneous. The natural way to construe that assertion is as a statement that since most, but not all, members of $S.F$ develop paresis, there must be some further factor G that helps to determine which members of $S.F$ will develop paresis (P) and which will not. If all members of $S.F.G$ develop paresis, then we have an objectively homogeneous reference class on account of the universal generalization $(x) [Sx.Fx.Gx \supset Px]$, but this is a case of trivial objective homogeneity. Moreover, the resulting explanation is no longer an I-S explanation, but rather D-N, for we have replaced our statistical law with a universal law. If it should happen, instead, that a very high percentage of members of $S.F.G$ develop paresis, but not all, then our explanation of paresis might be an improved I-S explanation, but it would still be epistemically relativized, since all I-S explanations are relativized. Again, the reason we have no I-S explanation simpliciter is that $S.F.G$ is not objectively homogeneous but only epistemically homogeneous. The natural supposition, once more, is that there is some as yet undiscovered factor H that helps to determine which members of $S.F.G$ will develop paresis and which ones will not.

The foregoing line of argument can be summarized, it seems to me, by saying that the thesis of epistemic relativization of I-S explanation implies that all I-S explanations are incomplete D-N explanations, and that in some vague sense, the adequacy of an I-S explanation can be measured by the degree to which it approximates a D-N explanation. The conclusion is that there are no bona fide I-S explanations per se, but only I-S approximations to bona fide D-N explanations. The relationship between I-S explanations and D-N explanations closely parallels the relationship between enthymemes and valid deductive arguments. Since an enthymeme is, by definition, an argument with missing premises—let us here ignore the cases in which the conclusion is suppressed—there can be no such thing as a valid enthymeme. Enthymemes can be made to approach validity, we might say, by supplying more and more of the missing premises, but the moment a set of premises sufficient for validity is furnished, the argument ceases to be an enthymeme and automatically becomes a valid deductive argument.

Much the same sort of thing can be said about I-S explanations. The reference class that occurs in a given I-S explanation and fulfills the requirement of maximal specificity is not genuinely homogeneous; it is still possible in principle to effect a relevant partition, but in our particular knowledge situation we do not happen to know how. As we accumulate further knowledge, we may be able to make further relevant partitions of our reference class, but as long as we fall short of universal laws, we have not exhausted all possible relevant information. Progress in constructing I-S explanations would thus seem to involve a process of closer and closer approximation to the D-N ideal. Failure to achieve this ideal would not be a result of the nonexistence of relevant factors sufficient to provide universal laws; failure to achieve D-N explanations can only result from our ignorance.[3]

As a result of the foregoing considerations, as well as other arguments advanced by J. A. Coffa in his penetrating article (1974a), I am inclined to conclude that Hempel's concept of *epistemic relativity of statistical explanation*, which demands relativization of *every* such explanation to a knowledge situation (Hempel, 1965, p. 402), means that his account of

---

[3] I. Niiniluoto pointed out that in infinite reference classes, it may be possible to construct infinite sequences of partitions that do not terminate in trivially homogeneous classes. It is clear that no such thing can happen in a finite reference class. This is, therefore, one of those important points at which the admitted idealization involved in the use of infinite probability sequences (reference classes) must be handled with care. The issue of finite versus infinite reference classes will be discussed in the final section of this chapter.

I-S explanation is completely parasitic upon the concept of D-N explanation. This conclusion resonates strongly with the Laplacian view, which we discussed at length in chapter 1. In the context of Laplacian determinism, it is reasonable to suppose that we will often have to settle for statistical explanations of various phenomena on account of our incomplete knowledge. Such explanations, however, would be partial explanations. The Laplacian demon would never have to settle for statistical explanations; it would be able to furnish a D-N explanation of any phenomenon whatever. The doctrine of essential epistemic relativization of inductive-statistical explanation thus seems to be an echo of Laplacian determinism. As I remarked in chapter 1, the Laplacian influence is insidious.

If, however, indeterminism is true—on any reasonable construal of that doctrine with which I am acquainted—some reference classes will be actually, genuinely, objectively homogeneous in cases where no universal generalization is possible. If radioactive decay is genuinely indeterministic, then the class of carbon 14 atoms is objectively homogeneous with respect to decay within the next 5,730 years. Half will decay within that period and half will remain intact, and there simply are no additional factors that can be used to make a relevant partition in that reference class. In view of this circumstance, it seems to me, we must have a full-blooded account of statistical explanation that embodies homogeneity of reference classes *not relativized* to any knowledge situation. I do not know whether indeterminism is true, though I think that we have strong physical reasons for supposing that it may be true. Regardless of whether indeterminism is true or false, we need an explication of scientific explanation that is neutral regarding that issue. Otherwise, we face the dilemma of either (a) ruling out indeterminism a priori or (b) holding that events are explainable only to the extent that they are fully determined. Neither alternative seems acceptable: (a) the truth or falsity of indeterminism is a matter of physical fact, not to be settled a priori, and (b) even if the correct interpretation of quantum mechanics is indeterministic, it still must be admitted to provide genuine scientific explanations of a wide variety of phenomena—indeed, it is doubtful that any other twentieth-century scientific theory can match quantum mechanics in explanatory power.

In (Salmon, 1974) I suggested, in the light of considerations such as the foregoing, that Hempel appears to be implicitly committed to some form of determinism because of his doctrine of essential epistemic relativization of inductive-statistical explanations. In private communications, however, he has informed me that he is not committed to any form of determinism. He has a different ground for maintaining the essential relativization of I-S explanations to knowledge situations.

(2) It might be argued that we can never be warranted in asserting the

objective homogeneity of the reference class mentioned in any statistical generalization because no such assertion can coherently be made—the very concept of objective homogeneity is not meaningful. This was Hempel's basis, as I understand him, for maintaining until recently the *essential* epistemic relativity thesis for I-S explanations. I find this reason for rejecting objective homogeneity suspect. One ground for this suspicion is the utterly straightforward character of objective inhomogeneity; namely, the existence of a partition that is statistically relevant to the attribute in question. It seems very strange indeed to maintain the meaninglessness of a concept whose negation is perfectly meaningful. Another ground for suspicion is the straightforward character of objective homogeneity in the trivial cases involved in universal generalizations. It begins to look as if the charge against nontrivial homogeneity is not incoherence but vacuity—which is precisely what objection (1) came to. If, however, we want to meet this second challenge to objective homogeneity in a thoroughly convincing fashion, it seems necessary to provide an adequate explication of that concept. Until this can be done, the development of an S-R basis conforming to the ontic conception of scientific explanation will rely upon an uncomfortably large promissory note.

Before undertaking this project, which will occupy the remainder of this chapter, I should like to make a few remarks on its significance. *First*, if the world is actually indeterministic, we seem to need the concept of objective homogeneity to describe that very indeterminacy. In a sample of radioactive substance composed of atoms of a single isotope, for example, some atoms undergo spontaneous radioactive decay within a certain time interval and others do not. If the indeterministic interpretation of quantum mechanics is correct, then there is no further characteristic of these atoms that is relevant to their decay within that time period. To formulate the thesis of indeterminism—whether it turns out ultimately to be true or false—we seem to need the concept of a homogeneous reference class, and this homogeneity must represent an objective feature of the real world, not merely a gap in our knowledge. *Second*, if indeterminism is true, then we have strong reason to believe that some of the basic laws of nature are irreducibly statistical. An irreducibly statistical law can be characterized, it seems to me, as a lawful statistical generalization of the form "$P(B|A) = p$" where some A are B, some A are not B, and A is objectively homogeneous with respect to attribute B. On the indeterministic interpretation of quantum mechanics, the laws of radioactive decay, as well as many other laws, are irreducibly statistical. We need a concept of objective homogeneity to capture the idea of an irreducibly statistical law.[4] *Third*,

[4] Railton (1978) relies heavily in his account of probabilistic explanation—mentioned in

if our philosophical theory of scientific explanation is to do full justice to twentieth-century science, it must furnish an autonomous pattern of statistical explanation. As I argued previously, the epistemic relativization of I-S explanations seems to make that mode of scientific explanation parasitic upon D-N explanation. One of the conspicuous features of twentieth-century physics, it seems to me, which distinguishes it from the classical theories of Newton and Maxwell, is the fact that it relies heavily—and unavoidably, I suspect—upon fundamental laws of the statistical variety. In (1977, p. 112), Hempel expressed the hope that a nonrelativized concept of statistical explanation can be developed, but he did not seem overly sanguine about the prospects of achieving that goal.

I shall now turn to the technical details of spelling out a concept of objective homogeneity. The task turns out to be more complicated than we might initially have imagined, and I do not have great confidence that the account given in the next two sections is entirely satisfactory. I do hope, however, that it represents some progress in the right direction. As we shall see in subsequent sections of this chapter, important philosophical issues hinge upon the concept of homogeneity, not only for those who adopt an ontic conception of scientific explanation, but also for those who prefer an epistemic conception.

RANDOMNESS

In introducing the concept of homogeneity for use in the S-R model, as a rough counterpart of Hempel's concept of maximal specificity, I rather casually remarked that it could be understood along lines similar to those employed by Richard von Mises (1957, chap. 1; 1964, chap. 1) to explain what he meant by a "place selection" (Salmon et al., 1971, pp. 40–51). At the time, I did not fully appreciate the importance of adopting the objective concept and avoiding the epistemically relativized concept in elaborating the S-R model of scientific explanation. Coffa's penetrating essay "Hempel's Ambiguity" (1974a) brought this point home with full force. In view of this consideration, I attempted a rigorous explication of objective homogeneity in "Objectively Homogeneous Reference Classes" (1977), but that treatment was seriously flawed in several respects. In the following presentation, I shall attempt to repair the damage and provide a more satisfactory account of this crucial concept.[5]

A reference class A is homogeneous with respect to an attribute B

---

footnote 1 of this chapter—on irreducibly probabilistic laws; beyond some sketchy hints, however, he does not tell us how they are to be characterized.

[5] I am especially indebted to Paul W. Humphreys for pointing out a number of difficulties and for discussing ways in which they might be circumvented.

provided there is no set of properties C_i ($1 \le i \le k; k \ge 2$) in terms of which A can be relevantly partitioned. By a partition of A we understand a set of mutually exclusive subclasses of A which, taken together, contain all members of A. A partition of A by means of C_i is relevant with respect to B if, for some values of i, $P(B|A.C_i) \ne P(B|A)$. To say that a reference class is homogeneous with respect to an attribute does not mean merely that we do not know how to effect a relevant partition, or that there is some practical obstacle to carrying out the partition. To say that a reference class is homogeneous—*objectively homogeneous* for emphasis—means that there is no way, even in principle, to effect a relevant partition.

There are two cases in which homogeneity obtains trivially, namely, if all A are B or if no A are B. This follows from the obvious logical theorems:

$(x) [Ax \supset Bx] \supset (x) [Ax.Cx \supset Bx]$
$(x) [Ax \supset {\sim}Bx] \supset (x) [Ax.Cx \supset {\sim}Bx]$

We shall not be interested in trivial homogeneity in the following discussion.

In the nontrivial cases, some restrictions must be imposed upon the types of partitions that are to be admitted; otherwise, the concept of homogeneity becomes vacuous in all but the trivial cases. Suppose that $P(B|A) = 1/2$. Let $C_1 = B$ and $C_2 = \bar{B}$. Then

$P(B|A.C_1) = 1;$
$P(B|A.C_2) = 0.$

A relevant partition has thereby been achieved.

The problem of ruling out unsuitable partitions is similar to the problem von Mises faced in attempting to characterize his *collectives*. A collective, it will be recalled, is an infinite sequence $x_1, x_2, \ldots$, in which some attribute B occurs with a relative frequency that converges to a limiting value p. Furthermore, the sequence must be *random* in the sense that the limiting frequency of B in any subsequence selected from the main sequence by means of a *place selection* must have the same value p. This is the principle of insensitivity of the probability to place selections; it is also the principle of the impossibility of a gambling system. Roughly speaking, a place selection must determine whether a member of the main sequence belongs to the subsequence without reference to whether the element in question has or lacks the attribute B. There are two types of place selections: (1) selections that determine membership in the subsequence entirely on the basis of the ordinal position of the element in the original sequence— for example, every third element, or every element whose place corresponds to a prime number—and (2) selections that determine the membership of the subsequence at least partly on the basis of attributes of members of the main sequence that precede the element in question—for example, every element that immediately follows two tails in succession in a sequence of coin tosses.

The definition has been challenged, on the ground that the concept of the collective, thus defined, is empty (except in the trivial cases). Given any sequence of elements A, each of which either has the attribute B or lacks it, there exists a real number between zero and one whose binary representation contains a 1 wherever the attribute B occurs and a 0 wherever B is absent in the sequence. This real number could thus furnish a place selection that would pick out every element in the sequence that has the attribute B and reject all that are not B. The original sequence would not be a collective, for we have shown that there exists a place selection with respect to which the limiting frequency of B is not invariant. The fact that we have no way of *knowing* in advance which real number would furnish a place selection—for a given sequence of coin tosses, for example—is irrelevant. As far as von Mises's original definition of the collective is concerned, all that matters is the *existence* of such a place selection.

An answer to this objection, which von Mises enthusiastically endorsed, was provided by Abraham Wald (see von Mises, 1964, pp. 39–43). It runs as follows. Given the obvious limitations of standard mathematical languages, at most a denumerable infinity of rules can be formulated in any particular language. If we limit the class of place selections to those that can be represented by explicit rules, then at most a small subset of the real numbers between zero and one can correspond to actually formulated place selections. If such a restriction to a denumerable set of place selections is imposed, the existence of nontrivial collectives is demonstrable.

This resolution of the difficulty is open to serious objection. Just as we must carefully distinguish between numbers and numerals (names of numbers)—noting that there is a superdenumerable infinity of real numbers but only a denumerable infinity of numerals—so also must we distinguish between the superdenumerable infinity of place selections that exist abstractly and the denumerable infinity of linguistic entities (rules or recipes) that represent them. Thus if the definition of "collective" rests upon the invariance of the limiting frequency with respect to the set of all place selections that exist, regardless of whether they are represented by explicitly formulated rules or not, then the concept of the collective remains empty. If, on the other hand, the collective is defined by reference to a set of formulated (or formulable) rules for effecting place selections, then the associated definition of "randomness" is relativized to a particular language in which the rules are to be formulated. The consequence is that a sequence that qualifies as a collective with respect to one language may fail to qualify as a collective with respect to another.

In view of this consideration, Hempel has, in private correspondence, rightly challenged my use of a concept of homogeneity explicated in terms of von Mises's notion of a place selection. His account of I-S explanation makes use, in effect, of a concept of homogeneity of reference classes that is relativized to knowledge situations. It is worth noting, in passing, that certain theories of probability (see, e.g., Kyburg, 1961) embody a concept of randomness that is likewise relativized to knowledge situations.[6] The S-R model of explanation, as presented in (Salmon et al., 1971), involved a concept of homogeneity defined in terms of von Mises's place selections; as such, it was relativized to a particular language. Relativization to a language—which might be construed as involving an entire conceptual framework—is, I believe, preferable to relativization to a highly ephemeral knowledge situation; indeed, some philosophers—for example, Fetzer (1981) and Tuomela (1981)—explicitly and deliberately construct models that are linguistically relativized. Each model involves some type of maximal specificity requirement that makes direct reference to a language. Inasmuch as both Fetzer and Tuomela regard explanations as arguments, it is hard to see, given the linguistic character of arguments, how relativization to a language could be avoided.

I am not really content with either epistemic or linguistic relativization. We need, it seems to me, a reasonable concept of homogeneity according to which a given reference class is *objectively homogeneous* with respect to the occurrence of a given attribute, quite independently of either the specific knowledge situation or any particular language. Such a concept is needed, even if we leave the question of scientific explanation aside, because it is involved in the idea of physical indeterminism. Moreover, if we relinquish the notion that explanations are arguments and adopt instead an ontic conception of scientific explanation, it is natural to characterize statistical explanations in a way that involves a nonrelativized concept of homogeneity.

It was evidently concern about the issue of language relativization, as well as worry about the Richard paradox, that led Alonzo Church (1940) to offer a refinement of the concept of the collective. Instead of defining place selections in terms of the rules that can be formulated explicitly in a given language, Church proposed to restrict them to selections given in terms of "effectively calculable" functions. As Church has defined this term, a function is effectively calculable if and only if it is λ-definable or (equivalently) it is general recursive. This concept has been shown by

---

[6] This seems to be a common feature of theories of logical probability; cf. (Carnap, 1950, pp. 493–495).

A. M. Turing (1937) to be equivalent to computability on a Turing machine. Let us call such selections *Church place selections*.

As Church has pointed out, his definition of "random sequence" has several advantages over various alternatives. (1) In contrast to von Mises's original definition, this one provides a concept that is demonstrably nonvacuous. It can be shown that, for each real number p ($0 \leqslant p \leqslant 1$), there exists an uncountable infinity of random sequences of 1 and 0 in which the limit of the frequency of 1s has that value p (Church, 1940). (2) Less stringent definitions, such as Copeland's admissible numbers (Copeland, 1928) or Reichenbach's normal sequences (Reichenbach, 1949), are too broad to serve as general concepts of randomness. According to these definitions, a sequence in which the limit of the frequency of 1 equals 1/2 could qualify as normal even if it had a 1 at each place corresponding to a prime number. Such concepts, though useful in certain contexts, are not sufficiently restrictive for the definition of homogeneity. They flagrantly violate the principle of the impossibility of a gambling system as well as insensitivity with respect to place selections. (3) The class of effectively calculable functions is well defined independently of any arbitrary choice of language. Church's random sequences have the property of randomness objectively, without relativization to any language or any knowledge situation. (4) Quite apart from any other considerations, it seems entirely reasonable to insist that place selections be defined in terms of effectively calculable functions. If someone directs us to select a subsequence from a probability sequence, it does not seem excessive to demand that there should exist, in principle, some method (or algorithm) by means of which it is possible to determine which elements of the original sequence belong in the selected subsequence and which ones do not.

Church's definition of "random sequence" does, of course, involve 'Church's thesis'—the claim that effective calculability coincides with the triad of mutually equivalent properties: λ-definability, general recursiveness, and Turing computability. While I realize that Church's thesis may be disputed, I am not aware of any reason for calling it into question in the present context.

The von Mises definition of randomness, even when modified so as to employ Church place selections, is not altogether without its problems. These have been reviewed by P. Martin-Löf (1969). Additional work by Martin-Löf (1966) and C. P. Schnorr (1971) appears to have overcome the difficulties. The subsequent developments are within the spirit of von Mises and Church; the requirement of effectiveness, with its language-independence, is retained throughout. In an excellent survey article, Coffa sums it up in the following way, "The end-result of this process seems to be a successful explication of the concept [of randomness] useful to the theoretically-minded statistician'' (1974, p. 107). If that goal has not been fully achieved, we can at least take comfort in the fact that the job is in very good hands.

HOMOGENEITY

The task of defining homogeneity is not yet finished, for this concept has an empirical as well as a mathematical aspect. Indeed, as we shall see, the empirical aspect is by far the more important of the two for our purposes. As Coffa (1974, p. 107) remarks in his very next sentence, ''The question remains whether [the new concept of mathematical randomness] relates in any interesting way to that of physical randomness.'' While I am satisfied that Church's earlier work—augmented by the more recent work of Martin-Löf, Schnorr, and others—has provided the means to deal with the mathematical aspect, we must say something about the empirical side, and how the two are related to one another.

The major problem, it seems to me, can be seen in connection with von Mises's original *informal* characterization of his desiderata. On the one hand, he spoke about insensitivity under place selections, which were defined entirely in terms of the *internal* properties of the sequence in question. Given an adequate definition of ''place selection''—of the sort discussed in the preceding section—and given a particular sequence, it is possible in principle to determine whether that sequence is random without reference to any other sequence of events. On the other hand, von Mises talked about the impossibility of a useful gambling system. It is true, of course, that various gambling systems have been proposed—such as martingales—that attempt to exploit the preceding outcomes of the sequence of trials in order to recoup losses and insure gains. Others—such as betting on every third member of the sequence—that depend only upon the place in the sequence and not upon previous outcomes are easily devised. Systems of both kinds are ineffective for gambling on any sequence that satisfies the condition of insensitivity to place selections of the type von Mises attempted to define. There is, however, the question of whether gambling systems of other types are possible in principle. Can there be any method for the placing of bets on events in sequences that fulfill all of the internal requirements for randomness and that will be guaranteed to win? As the following simple example shows, the answer is yes.

EXAMPLE 1: Suppose a dishonest roulette wheel is rigged in such a way that the croupier can determine in advance whether any given spin of the wheel will result in red or black. In order to be sure that the sequence of outcomes appears to be entirely random, the croupier has a confederate who tosses a penny before each spin of the wheel. The confederate secretly signals the result to the croupier. If the penny lands heads up, the croupier makes the roulette wheel produce the outcome red; if the penny lands tails up, the croupier arranges for the outcome on the wheel to be black. It is obvious that anyone who knew of this arrangement would have available a splendid gambling system—one that would guarantee a win on every trial. Assuming that the penny produces a random sequence of heads and tails, the sequence of red/black roulette outcomes would obviously possess all of the internal characteristics of randomness of the mathematical variety.

Under the circumstances outlined in this example, it seems clear that we should not classify the sequence of spins of the roulette wheel as homogeneous with respect to the properties red/black. In this case, it is a relationship between the two sequences—the flips of the penny and the turns of the roulette wheel—rather than any internal property of one sequence, which makes us deny that it constitutes a homogeneous reference class. It is strange, in my opinion, that von Mises, who often made reference to the impossibility of gambling systems, never dealt with cases of this sort. This is precisely the kind of situation with which we must deal in order to arrive at a satisfactory characterization of objective homogeneity.

The best way to approach this problem, it seems to me, is to add to von Mises's requirement of *invariance of the limiting frequency with respect to place selections* a requirement of *invariance of the limiting frequency with respect to selections by associated sequences*. The formulation of this requirement will involve several crucial definitions. The first of these is fairly easy.

DEFINITION 1: *Let A be a reference class consisting of a sequence of events x_1, x_2, . . . Any other infinite sequence D consisting of events y_1, y_2, . . . will be called an* associated *sequence if each event y_i occurs in the absolute past (past light cone) of the corresponding event x_i.*

While this definition rules out the possibility that x_i and y_i are one and the same event, it does not exclude such possibilities as $y_i = x_{i-1}$ (i.e., y_i is simply the immediate predecessor of x_i in A).

As usual, let B be an attribute class whose probability within the sequence A concerns us.[7] Let C be an attribute class to which any y_i in D may belong or not—that is, for every i, either $y_i \in C$ or $y_i \in \bar{C}$. We want to use C to

[7] In this discussion, I shall use terms like "class," "attribute," and "property" more or less interchangeably. This practice will have stylistic advantages, and there are no particular dangers involved, since we shall not be dealing with the kinds of classes and properties that give rise to logical or set-theoretical paradoxes.

define a *selection by an associated sequence*. We shall define a selection S from the reference class A by means of the associated sequence D by stipulating that $x_i \in S$ *iff* $y_i \in C$. The subscript i on the members of A and D establishes a one-one correspondence between the members of these two ordered classes. In example 1, the class A consists of the spins of the roulette wheel, and the attribute B is the outcome red. The associated sequence D consists of the flips of the penny, and the attribute C is the outcome heads. The selection S is the sequence of spins of the roulette wheel immediately following tosses on which the penny lands heads up. Each spin of the roulette wheel x_i is associated with a corresponding toss of the penny y_i, and each y_i is in the absolute past of the x_i to which it corresponds. The probability of B within A is obviously *not invariant* with respect to the selection S.

Using this new concept—which we have introduced informally but not defined as yet—we shall want to say that A is not homogeneous with respect to B if there exists a selection S by an associated sequence such that the probability of B within $A.S$ is not equal to its probability within A—in other words, the reference class A is homogeneous with respect to B only if the occurrence of C within D is statistically irrelevant to the occurrence of B within A. This, in turn, is tantamount to the requirement that the sequence of Bs within A be statistically independent of the sequence of Cs within D. In order to avoid making the concept of homogeneity vacuous—except in the trivial cases—it is necessary to impose certain restrictions upon the class C, or equivalently, upon the properties that determine the membership of C. Coffa (1974) called attention to this problem in his discussion of *physical properties*, and I discussed them at length in (Salmon, 1977). Nevertheless, in his critique of the S-R model, James Fetzer (1981, pp. 86–103) appeals to just such trivializing moves in an attempt to show that the S-R model entails determinism. His argument is vitiated by his failure to take notice of the kinds of restrictions to which Coffa and I had called attention. In the discussion that is to follow, I shall try to provide an improved account of the conditions that must be fulfilled by any property that is to be used to define an ''admissible selective class.''

Although we are not primarily interested in the epistemic or practical aspects of gambling, the notion of a possible gambling system has, as von Mises realized, considerable heuristic value. In order to have a workable gambling system, one obviously has to be able to decide, before the occurrence of the event on which one is going to bet, how to bet on it. Thus if a selection by an associated sequence is to provide a usable gambling system, we must be able *in principle* to know whether the event y_i has the property C before the event x_i occurs. This fact imposes conditions both upon the events y_i and upon the property C. We have already secured the basic condition upon the events y_i by requiring, in definition 1, that each y_i temporally precede the corresponding x_i. This requirement, which seems so obvious, has important applications in dealing with the concept of homogeneity, as we shall see later in this chapter. The appropriate restriction upon C is much harder to secure. Roughly speaking, we must be able in principle to ascertain whether y_i has or lacks property C before x_i occurs. As the following example shows, the fact that y_i must occur before x_i does not, in itself, guarantee that C will satisfy this condition.

EXAMPLE 2: Let A be a sequence of weather conditions on successive days in New York City, and let D be a sequence of forecasts, each of which is broadcast on the six o'clock news on one of the city's television channels on the preceding evening. Let B be the occurrence of a storm in New York, and let C be a reliable forecast of a storm. For purposes of this example, we shall take a *reliable* forecast to be, by definition, a prediction that comes true.

Since each forecast y_i occurs on the day before the corresponding weather condition x_i, this example does not violate our definition of an associated sequence. Clearly, however, it involves an illegitimate property C. The problem is that, as a logical consequence of our definitions of B and C, $y_i \in C$ if and only if $x_i \in B$. We must require of any property C that is to be used to make a selection by an associated sequence that it be defined in such a way that y_i's possession of C is logically independent of whether x_i has property B. One could not use the occurrence of a reliable forecast as a basis for betting on tomorrow's weather, for one would have to wait until tomorrow to find out whether the forecast is a reliable one.

There is a further problem concerning restrictions upon the property C that must be addressed. It is best introduced by means of another example:

EXAMPLE 3: Let A be a sequence of days, and let B be days on which there is a fatal accident in the town of Centerville. Let us assume that Centerville has a daily paper, the *Centerville Gazette*, which reports with reasonable accuracy on the fatal accidents that occur in that town on the preceding day. Note that in this example—in contrast to example 2—the concept *report of a fatal accident* is logically independent of the concept *occurrence of a fatal accident*, for neither logically entails the other. Let D be the sequence of daily editions of the paper. We establish the correspondence between the two sequences A and D by coordinating each issue of the paper with the succeeding day—for example, if x_i is July 4, then y_i is the issue of the paper that appears on July 3. Let the attribute C be the property of *carrying a dateline two days earlier than that of an issue containing a report of a fatal accident*. Thus if the July

5 issue of the *Gazette* contains a report of a fatal accident in Centerville, then the July 3 issue of the paper has property C.

In this example, y_i is in the absolute past of x_i, so D does qualify as an associated sequence. Moreover, there is no *logical* connection between the fact that a given y_i has (or lacks) property C and the fact that the corresponding x_i has (or lacks) property B. We may suppose, for purposes of this example, that not every fatal accident is reported and that reports of accidents occasionally state erroneously that a fatality occurred; it is sufficient to have a high correlation between reports of fatal accidents in the *Gazette* and the occurrence of fatal accidents on the preceding day. Nevertheless, we would not want to allow a selection S based upon the property C in example 3 to have any bearing upon the homogeneity of A with respect to B. The reason is that although strictly speaking C applies to events that occur before the associated members of A, it is essentially defined in terms of events that occur after the fact—namely, newspaper reports of fatal accidents. The relationship to a possible gambling system is obvious. We are not allowed to bet on events after they have occurred, at least if there is any possibility of our receiving a report on the outcome before placing the bet. A bookie who took bets from people who had already heard radio reports on the outcomes of races would not be in business very long. If, in example 3, we knew which members of D had attribute C, we would be able to use that information to make a relevant partition in A with respect to fatal accidents in Centerville. We cannot, of course, use this partition for purposes of prediction, for we cannot know which members of D do possess attribute C until it is too late to make predictions.

It might seem that we are now in a position to formulate with reasonable rigor the restrictions that must be placed upon attribute C if it is to be used to define "selection by an associated sequence." Any C that satisfies appropriate restrictions will be called an "admissible selective class" (or "attribute" or "property"). In order for a given C to qualify, it must be possible in principle to ascertain whether y_i has attribute C before event x_i occurs. Moreover, whether y_i has C must be logically independent of the properties of the corresponding x_i as well as the properties of events that are subsequent to x_i. C must not be defined in terms of properties of x_i or later events. Unfortunately, an adequate definition of "admissible selective class" is not quite that simple.

In the first place, the concept of a property of an event must be used with extreme caution. The July 3 issue of the *Centerville Gazette* does have the property of preceding by two days an issue that contains a report of a fatal accident. It is a property of the Fourth of July accident to be reported in the July 5 issue of the *Gazette*, and to occur the day following an issue of the paper that carries a dateline two days earlier than an issue that contains a report of a fatal accident. It is a property of the July 5 issue of the newspaper to be preceded by the July 3 issue. Evidently, there are serious problems about the localization of properties.

In the second place, we must recall what we are attempting to do, namely, to characterize *objective* homogeneity. We are trying to avoid both epistemic relativization and linguistic relativization. But we have been talking freely, for heuristic purposes, about gambling and about what we can and cannot know in attempting to devise a gambling system. To escape epistemic relativization, we must avoid using such epistemic notions as knowledge and prediction in framing our restrictions. Similarly, to escape linguistic relativization, we must refrain from using such linguistically dependent concepts as logical dependence or independence, for logical entailment is a language-dependent concept. In addition, we must eliminate reference to definition in these restrictions, for definition is also a linguistically relative concept. We have been using these concepts without much restraint.

The problem we are facing was addressed, in a somewhat similar context, by Hans Reichenbach. He introduced the concept of a "codefined class," according to which, "we say that a class A is *codefined* if it is possible to classify an event x as belonging to A coincidentally with the occurrence of x . . . observing x we must be able to say whether x belongs to A, and it must be unnecessary to know, for purposes of this classification, whether certain other events y, z, . . . occurred earlier or later, or simultaneously at distant places" (1956, p. 187). Informally, Reichenbach's intent seems plain enough. He elaborates: "Logically speaking, a codefined class term is a one-place predicate which is not contracted from many-term predicates" (ibid.).

Reichenbach's remarks about codefined classes can hardly be taken to provide a precise characterization of that concept. First, those who have worried about Nelson Goodman's predicates "grue" and "bleen" will naturally wonder whether they enjoy the status of uncontracted one-place predicates. I think that a negative answer to that question can be established, for it follows from a resolution of Goodman's puzzle that I offered in (Salmon, 1963), but I shall not attempt to reargue that issue here. Second, Reichenbach seems to suggest that codefined classes must be determined by directly observable properties. Many philosophers, nowadays, deny the viability of any sharp distinction between the observational and the theoretical (or more properly, unobservable). It seem to me that, whatever stand one takes on that issue, there is no reason to exclude from the realm of codefined classes those that are determined by unobservable properties. The class of spontaneous radioactive decays consists of events that are not directly observable, but I think it should qualify as codefined. What seems essential to the concept of a codefined class is that the inclusion or exclusion of events should be determined by the spatiotemporally local characteristics of the events involved. As we remarked previously, however, it is not totally clear just how the locality of *characteristics* is to be defined.

I shall assume that the concept of the location of an *event* in space-time is clear enough. We know how to delineate approximately the space-time region in which a coin toss, a thunderstorm, a radioactive decay, or a supernova explosion occurs. Thus the events x and y that constitute the membership of classes A and D are taken to have definite space-time locations (at least to a reasonable approximation). When Reichenbach stipulates that the classes A, B, C, and D be codefined, he means, roughly speaking, that it is possible in principle to ascertain whether a given event x belongs to one of these classes by examining the space-time region in which x occurs. This rough characterization suffers, however, from the fact that it is framed in epistemic terms. We want to say, nonepistemically, that the membership of x in any of these classes is objectively determined by facts that obtain within the space-time region in which x occurs. This statement does not help us much, for, as we noted, it seems to be a fact about the issue of the *Centerville Gazette* that appears on July 3 that it carries a dateline two days earlier than the dateline on an issue containing the report of a fatal accident. Moreover, Reichenbach's attempt to clarify the concept of codefined classes by using the notion of an uncontracted *predicate* introduces just the sort of linguistic relativity we were concerned to avoid.

There are two main steps that can be taken to adapt Reichenbach's definition to our purposes. First, we can replace his human observer with a physically possible detector, and his human knower with an electronic computer. The physical detecting apparatus may be any sort of device that responds, for example, to a pulse of light (a photocell), the impact of a material particle (a ballistic pendulum), or the presence of a metallic object (a magnetometer). Any such detector is admissible, provided only that it is physically possible—that is, its existence would not be excluded by any law of nature. This detector is connected to a computer. When the event y_i occurs, the computer receives a message from the detector. If y_i has the property F to which the detector responds, then it sends the message "y_i-yes"; if not, it sends the message "y_i-no."

The property F may or may not be identical to the property C that we intend to use to make a selection. If they are identical, we shall want to say simply that the detector must be able to respond to y_i before the associated event x_i occurs. In most cases—that is, those in which x_i does not follow too closely on the heels of y_i—it will be all right to require that the detecting apparatus transmit the result before x_i occurs.

It frequently happens in scientific investigations that the physical detector responds to some property F other than the property C we are trying to measure. For instance, in the famous 1919 test of general relativity, Eddington endeavored to ascertain the amount of bending experienced by light passing close to the sun; what he actually measured was the distance between spots on a photographic plate. From the results of these measurements he *calculated* the deflection. Because this sort of case is typical, I suggested that the detector be attached to a calculator. The computer would accept the result concerning y_i's possession of F and would proceed to tell us whether y_i belongs to C or not. It would be unreasonable to demand that the calculator produce the answer before the corresponding x_i occurs, but we can require that the computer receive no further inputs of data until it has furnished that answer. It would be impossible in principle for a detector/computer of this sort to ascertain whether a given weather forecast is reliable (example 2), or whether a given issue of the *Centerville Gazette* has the property of preceding by two days an issue that carries a report of a fatal accident (example 3). We may consider a class C *objectively codefined* if it is physically possible to have a detector/computer of the foregoing sort to ascertain its membership. Defined in this way, the concept of an objectively codefined class is neither epistemically nor linguistically relativized.

There is a second reformulation of Reichenbach's remarks on codefined classes that may have some value in dealing with the problem about admissible selective classes. Consider any statement of the form "$y_i \in C$ if and only if $z_i \in F$." Such statements are to be construed in terms of the following stipulation:

$y_i \in C$ iff z_i occurs and $z_i \in F$
$y_i \in \bar{C}$ iff z_i occurs and $z_i \in \bar{F}$, or z_i does not occur.

Let us apply this stipulation to example 3. If the July 5 edition of the *Centerville Gazette* appears and carries a report of a fatal accident, then the July 3 edition belongs to the class C of issues carrying a dateline two days earlier than an edition that carries a report of a fatal accident. If the July 5 edition comes out and does not contain a report of a fatal accident, or if the July 5 edition does not appear at all, then the July 3 edition is excluded from that class. Let us suppose for the sake of example that the July 5 edition does appear, and it carries a report of a fatal accident. We see that the membership of the July 3 edition would be switched from C to $\bar{C}$ if an event in the future—namely, the appearance of the July 5 edition of the paper—were to fail to occur. Classes whose membership can be altered in that way should not be allowed for purposes of making selections. This is our nonepistemic and nonlinguistic reformulation of Reichenbach's demand that we must be able to classify y_i as a member or nonmember of C regardless of the occurrence of other events at other places and times. Our restriction is, however, less stringent than Reichenbach's in one respect. We shall not exclude events in the absolute past of y_i from having a role in determining whether or not y_i belongs to C, for news of such occurrences could reach our detector/computer apparatus soon enough to be taken into account in making any calculation. Let us enshrine these considerations in an official definition.

DEFINITION 2: *The class C is an* objectively codefined subclass *of D if and only if, for any $y_i \in D$, (1) the membership of y_i in C or $\bar{C}$ could, in principle, be ascertained by a computer that receives information from a physical detector responding to y_i, but that receives no information gathered by the detector (or from any other source) in response to x_i or to any events z_i in the absolute future of x_i, and (2) there is no event $y_i \in D$ whose membership in C or $\bar{C}$ would be altered if x_i or any event z_i in the absolute future of x_i were not to occur.*

The foregoing definition succeeds, I believe, in blocking the kinds of properties—illustrated in examples 2 and 3—that illicitly make use of facts in the future of y_i to determine whether it belongs to C. It seems likely that clauses (1) and (2) are somewhat redundant with one another; perhaps one of them would be sufficient by itself. Both are given because each of them formulates an important intuition, and in dealing with an issue as subtle as objective homogeneity, a bit of overkill may not be a bad thing.

Before attempting to arrive at precise formulations of the restrictions that must be imposed in order to characterize admissible selective classes, we must consider one additional problem. It harks back to the original difficulty (discussed in the preceding section) that bedeviled von Mises's definition of the collective. It crops up again, as example 4 will show, in connection with associated sequences.

EXAMPLE 4: Consider a superstitious crapshooter who invariably says, just before rolling the dice, "Gimme a ———," where the blank is filled by "seven" or the name of whatever number he hopes to get. Let us consider only those tosses in which he is trying for 7. Although he thinks this improves his chances of getting 7, it does not, in fact, make any difference at all. The chance of his rolling a 7 is 1/6, just as it would be if he did not say the 'magic words.' Let A be the (potentially infinite) sequence of his rolls of the dice, and let B stand for the result being 7. Let D be the sequence of his utterances of the words "Gimme a seven" just before throwing the dice. D qualifies as an associated sequence, for each of its members precedes the element of A to which it is coordinated. We define the attribute C as follows. The outcomes of the tosses of the dice can be represented by a sequence of 1s and 0s where each case in which the result is 7 is represented by a 1 and each of the others is represented by a 0. If the sequence of 1s and 0s is preceded by a dot, it becomes the expression in binary notation for some real number r between zero and one. Now, let the ith utterance y_i have the property C if and only if the ith numeral in the binary expansion of r is a 1. If we use this attribute C to make a selection S in A, then S will pick out all and only those tosses that result in 7.

It is clear that selections of this sort have no bearing upon the question of whether A is homogeneous with respect to B. If we were to allow them to be used, homogeneity would become a vacuous concept (except for the trivial cases). As we remarked in connection with the original objection to von Mises, it does no good to protest that we have no way of knowing what number r will do the trick. The point is simply that some such number exists, and consequently the selection S defined in example 4 also exists. In the present context, the point is even more obvious, for we are explicitly seeking to characterize a concept of homogeneity that is neither epistemically nor linguistically relativized. We will have to exclude properties of this sort when we attempt to impose appropriate restrictions upon the kinds of properties C that can be used to make selections in A.

The most straightforward way of dealing with this problem is, I believe, to use the method that successfully handled the same problem for the concept of mathematical randomness. Indeed, we can simply import that result by using the concept of mathematical randomness in the present context, as follows:

DEFINITION 3: *The class C qualifies as an* admissible selective class *if and only if (1) C is an objectively codefined subclass of D (as detailed in definition 2) and (2) C occurs within D in a mathematically random fashion.*

Clause (2) of this definition is designed to prevent us from importing, through the use of associated sequences, the kinds of selections that had to be ruled out in order to arrive at a satisfactory definition of mathematical randomness. The class C introduced in example 4 will be blocked because the binary numerical expression used to define that class will be noncomputable if (as we are assuming) 7s occur randomly among the rolls of the dice. This result prepares the way for our next definition:

DEFINITION 4: *Let the ordered class y_1, y_2, . . . constitute an associated sequence D with respect to the reference class A (as detailed in* *definition 1). Then, a* selection by an associated sequence S *is any selection within A defined by the rule*

$$x_i \in S \text{ iff } y_i \in C,$$

where C is an admissible selective class (as detailed in definition 3).

The motivation for this definition is the need to impose, as a condition on the homogeneity of a reference class, a requirement of invariance of the frequency of the attribute in question under any such selection by an associated sequence. Definition 4, which seems to embody all of the restrictions of which mention has been made, can be used to formulate the definition that was the goal of our discussion:

DEFINITION 5: *A reference class A is* objectively homogeneous *with respect to an attribute B iff the probability of B within A is invariant under all selections by associated sequences.*

It should be noted that this definition precludes reference classes A in which B occurs nonrandomly from being considered homogeneous, for any place selection S within A can easily be used to generate a selection by an associated sequence.

In view of the importance of the relation between objective homogeneity and indeterminism, let us apply our definitions to an example in which there is some plausibility to the claim that a nontrivial, objectively homogeneous reference class is involved.

EXAMPLE 5: Suppose that we have a sample of some radioactive material; this sample consists of atoms of one particular isotope of one particular element. Geiger counters are so situated as to detect any radioactive decay that occurs in that sample. Assume, further, that these detectors are connected with a tape recorder that records a "click" on a magnetic tape whenever one of the counters detects a decay-event. Assume also that the speed of the tape across the recording head is one cm/sec. Now, let A be the sequence of seconds during which this setup is in operation, and let B be the class of seconds during which at least one decay-event occurs. Let us stipulate that the sample is small enough and the nuclei are stable enough to yield a nonvanishing probability of seconds during which no decay occurs in the sample; indeed, for simplicity, let the probability for at least one decay during any given second be 1/2. Let D be the sequence of passages of centimeter-long segments of the tape across the recording head; some of the members of D will contain the clicks and others will not. We do not need to assume that the counters are perfectly reliable, or that the recorder is faultless; it is sufficient to assume a fair degree of reliability in each case. Now, using a definition quite parallel to that employed in example 3, we define C

as the class of passages over the recording head of centimeter-long segments that immediately precede those segments containing at least one recorded click (see Fig. 1). While there is some delay between the occurrence of a decay and its recording on the tape, we can assume that it is negligible in comparison with the second-long durations we are considering.

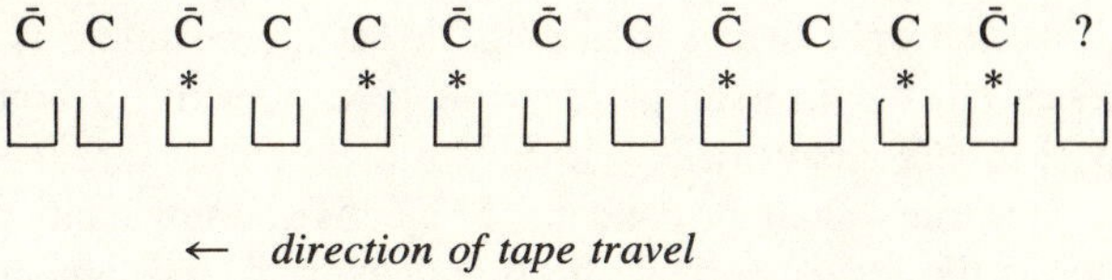

$$\bar{C} \quad C \quad \bar{C} \quad C \quad C \quad \bar{C} \quad \bar{C} \quad C \quad \bar{C} \quad C \quad C \quad \bar{C} \quad ?$$

| | | * | | * | * | | | * | | * | * | |

⟵ *direction of tape travel*

An asterisk in a segment signifies at least one click.

FIGURE 1

In this example, as in example 3, each y_i is in the absolute past of the corresponding x_i; therefore, D constitutes a bona fide associated sequence. Again, there is no *logical* connection between a given x_i possessing the attribute B and the corresponding y_i possessing the attribute C. Nevertheless, when our official definition of an admissible selective class is applied, we see immediately that C is disqualified. There is no physically possible detector that can detect, during second $n - 1$, a click that occurs during second n; likewise, there is no detector that can detect a decay-event in the second before it occurs. Moreover, according to the most widely accepted theory, there is no other physical fact that can be detected, during the second prior to a radioactive decay, that is statistically relevant to the occurrence of a radioactive decay-event, and consequently, none that is statistically relevant to the occurrence of a click during the succeeding second. Thus our definition rules out the class C in example 5.

Suppose someone were to suggest, in connection with example 5, that a different attribute C' might undermine the presumption that our original reference class A in that example is objectively homogeneous. The attribute C' would be defined along lines suggested in example 4. Consider a number r having the property that its binary expansion contains a 1 in the nth place if and only if a radioactive decay occurs in the nth second of our experiment. Then, the passage of the nth centimeter of our tape has the attribute C' if and only if r has a 1 in its nth place. If we idealize our example by supposing that it can go on forever, we will have an infinite sequence of digits. The class C' defined in this way will not, however, qualify as an admissible selective class. According to our best physical theories, the sequence of radioactive decays is a random one; the corresponding number r will consequently turn out to be an uncomputable number.

This is not the place to go into a detailed discussion of the completeness of quantum theory, but a brief remark about the bearing of this issue on objective homogeneity is in order. A common feature of many textbook presentations of quantum mechanics is the inclusion of a basic postulate to the effect that the quantum mechanical description of a physical system is complete.[8] Physical systems simply do not possess physical properties in addition to those embodied in the ψ-function. I do not want to insist that this postulate is true—though I think we have good reasons for believing it is—but only that it is intelligible. If the postulate is true, then the reference class A in example 5 is, in fact, objectively homogeneous with respect to radioactive decay. If the postulate makes sense—regardless of whether it is true—then the concept of objective homogeneity also makes sense. This is the result we have been endeavoring to establish.

Example 5 exhibits a deep physical fact about the world that has a crucial bearing upon the concept of objective homogeneity. Let us grant—for the sake of argument, at least—that spontaneous radioactive decay is an ineluctably statistical phenomenon. This means that for atoms of a particular isotope, there is no *prior* physical fact that is statistically relevant to the occurrence of a radioactive decay. No prior characteristic can be used to effect a relevant partition in the class of atoms of that species. After the fact, however, there may be a record of the chance event that enables us to know with a high degree of reliability that such an event has occurred. Although there is nothing logically impossible in the supposition that there might be records of future events as well as past events, in our world, I feel quite sure, there are records of the past but none of the future. In a world radically dissimilar to our world—one which did not exhibit the kinds of temporal anisotropies that we find in this world—the problem of defining objective homogeneity would take quite a different form. The characterization I am offering, even if it were totally successful, would not apply in all possible worlds.

SOME PHILOSOPHICAL APPLICATIONS

As we saw in the first section of this chapter, Hempel imposed a requirement of maximal specificity upon his inductive-statistical pattern of scientific explanation. Although it is epistemically relativized, it is designed

---

[8] White (1966, p. 30) provides a fairly typical formulation: "POSTULATE 3: *For every dynamical system there exists a state function ψ which contains all the information that is known about the system.*" Although this statement seems to refer to a possibly limited knowledge situation, the development of the theory makes it clear that the state function contains all possible information about the system. In (Leighton, 1959), another standard text, the corresponding postulate (pp. 93–94) says explicitly that the state functions "completely define the state of the quantum-mechanical system."

to fulfill the same function within his theory of explanation as our concept of objective homogeneity is intended to fulfill within the statistical-relevance approach. Requirements of these sorts are intended to block 'explanations' that omit relevant considerations. It seems to me, however, that even within Hempel's epistemically relativized theory, his requirement of maximal specificity does not do the job it is supposed to do, for reasons closely connected with our discussion of objective homogeneity.

Let us look at Hempel's formulation of that requirement:

Consider a proposed explanation of the basic statistical form

$$p(G,F) = r$$

(3o) Fb
$$\underline{\qquad\qquad} [r]$$
Gb

Let s be the conjunction of the premises, and, if K is the set of all statements accepted at the given time, let k be a sentence that is logically equivalent to K (in the sense that k is implied by K and in turn implies every sentence in K). Then, to be rationally acceptable in the knowledge situation represented by K, the proposed explanation (3o) must meet the following condition (the requirement of maximal specificity): If $s \cdot k$ implies[20] that b belongs to a class F_1, and that F_1 is a subclass of F, then $s \cdot k$ must also imply a statement specifying the statistical probability of G in F_1, say,

$$p(G,F_1) = r_1.$$

Here, r_1 must equal r unless the probability statement just cited is simply a theorem of mathematical probability theory. [Footnote 20 points out that this requirement must obtain even in cases in which the explanans s is not entirely contained in our body of accepted knowledge; hence the reference to $s \cdot k$ instead of k alone.] (1965, pp. 399–400).

In his (1968), Hempel makes certain modifications in the requirement of maximal specificity, but none of them affects the issue I am about to raise in any way.

As Hempel points out, the "unless" clause in the last quoted sentence is needed to block the danger of making the concept of a maximally specific class vacuous (except in the trivial case), for the class $F.G$ is always a subclass of F, and $p(G,F.G) = 1$ is a trivial theorem of the probability calculus. Thus, according to Hempel's requirement, if $F_1 = F.G$, the fact that $r \neq r_1$ is no obstacle to the maximal specificity of F. Unfortunately, even for an epistemically relativized concept of maximal specificity, Hempel's "unless" clause is not strong enough to protect against vacuity, as the following example shows.

EXAMPLE 6: Let A be a sequence of draws of balls from an urn, and let D be precisely the same sequence of draws—that is, $x_i = y_i$. Let B be the attribute of being a draw that results in the selection of a red ball. Suppose that $P(B|A) = 1/3$. Let C be the attribute of being a draw that results in the selection of a ball whose color is at the opposite end of the visible spectrum from violet. Evidently, $P(B|A.C) = 1$. In this example, I am assuming that it is *not* a logical truth—but, rather, a physical regularity—that red is at the opposite end of the visible spectrum from violet.

In (Salmon et al., 1971, pp. 49–50), I offered precisely this example to show that Hempel's requirement of maximal specificity is not adequately formulated; in his (1977, pp. 106–107), Hempel responded by saying that he considers the reference class A in example 6 to be inhomogeneous. This response is unsatisfactory, for if such answers were permitted in a general way, the concept of homogeneity would be trivialized. As I shall show shortly, it is always possible to use the same trick to show any reference class—other than the trivially homogeneous classes that occur as subject terms in universal generalizations—to be inhomogeneous. Raimo Tuomela (1981, pp. 266–267) saw clearly the point of such examples as 6, and he sought to block them by offering a revision of Hempel's requirement of maximal specificity, which he designates RMS_0. This revised version expands the "unless" clause of Hempel's condition that "r_1 must equal r unless the probability statement just cited $[p(G,F_1) = r_1]$ is simply a theorem of mathematical probability theory" to say, in effect, "unless the probability statement just cited is a theorem of mathematical probability theory or is a consequence of a law of nature."

Tuomela's revised maximal specificity requirement does not, however, provide an adequate means for dealing with other sorts of cases. Its adoption would have the effect of rendering nonrandom (inhomogeneous) all sorts of sequences that—for many purposes at least—are regarded as random. It is a contingent fact—not a consequence of the laws of logic or the laws of nature—that the side 6 of a standard die is marked with two parallel rows of three dots each, rather than with the roman numeral VI. Thus we could let B be the landing of a die with face 6 uppermost and C its landing with a face embossed with two parallel rows of three dots each on top. C could then be used to define a selection that would pick out all and only those tosses resulting in 6. Thus if Hempel's response to example 6 is correct, then a sequence of tosses of a standard die in the standard manner would not be homogeneous with respect to the result 6. Tuomela's reformulation does not help in this case. Similarly, I take it as a contingent fact, easily ascertained before tossing, that the Australian two-cent piece I am about to flip has a likeness of Elizabeth II on the side designated as heads. Again, we can let B be the attribute of heads showing and C the attribute of a likeness of Elizabeth II showing. C could then be used to define a selection that would pick out all and only those tosses of this particular coin that result in heads, thus rendering the class of standard tosses of a fair coin inhomogeneous with respect to getting heads. It is easy to see how this sort of ploy could be used in connection with any kind of 'chance device' to deprive it of its random character.

It can be argued, of course, that such sequences as coin tosses are not *really* random because of the fact that with sufficiently detailed knowledge of the state of the coin just before it lands, we would be able to predict with great accuracy which tosses will result in heads and which in tails. Such a claim may well be true. If, however, the class of tosses is inhomogeneous with respect to the outcome heads, it is because it can be partitioned in terms of the state of the coin *prior to* the result, not because every case of a head showing is a case of a likeness of the Queen being on the upper side. It is for this reason that the definition of an associated sequence includes the stipulation that the associated event must be in the absolute past of the member of the original sequence to which it is coordinated. Such a restriction seems necessary, even for epistemic homogeneity, if the concept of homogeneity is not to prove vacuous except in connection with universal generalizations. Without some such restriction, there could not be any scientific explanations that conform to Hempel's I-S pattern.

The same consideration enables us to deal with an example offered by John L. King (1976) as an objection to earlier formulations of the S-R model of scientific explanation.

EXAMPLE 7: Suppose, slightly simplifying the actual situation, that the neutral K-meson K^0 can decay in either of two ways: in mode N (neutral) it decays into a pair of neutral pions; in mode C (charged) its decay products include a positive pion and a negative pion. Among the class K of all K^0 decay-events, approximately 25% are of mode N and the remaining 75% are of mode C. We would want to say, presumably, that the reference class K is homogeneous with respect to spontaneous decay; however, it can be relevantly partitioned. Let P (plus) be the class of decays yielding a positive pion, and let M (minus) be the class of decays yielding a negative pion. Each and every member of K that has the attribute P also has the attribute M and conversely; this is not because of any logical relationship between P and M, but, rather, because of the physical law of conservation of charge. Now, the probability that a K^o decay will yield a negative pion—$P(M|K) = 0.75$—is not equal to the probability that a K^o decay that yields a positive pion will yield a negative pion—$P(M|K.P) = 1$. It follows that K is not homogeneous with respect to M.

This conclusion is not acceptable. Using our definitions of objective homogeneity and related concepts, which were not available to King when he published this example, we can escape the unwanted conclusion (see Thomas, 1979). The question we must consider is whether the occurrence of M is invariant with respect to associated sequences, as required by definition 5. As far as King's example is concerned, we need not give a negative answer. When we try to set up the associated sequence, we immediately find ourselves blocked by definition 1. Our basic sequence $x_1, x_2, \ldots$ is the class K of K^o decays, and the attribute with which we are concerned is M. We attempt to set up another sequence $y_1, y_2, \ldots$ in which the y_is will be associated with the corresponding x_is, and in which the attribute used for selection will be P. The difficulty is that the sequences are not distinct from one another; for every i, $x_i = y_i$, and so the y_is are not located in the past light cones of the corresponding x_is. Thus one and the same event has the attribute M and the attribute P, and we have not succeeded in setting up an associated sequence conforming to definition 1. Example 7, like example 6, which it closely resembles, shows that considerable care must be exercised in setting up selections by associated sequences.

Finally, a brief note should be added concerning the arguments adduced by Fetzer (1981, pp. 92–93) to show that my concept of homogeneity leads to the conclusion that the only homogeneous reference classes are those occurring as subject terms in universal generalizations. These arguments are based upon the mere existence of various subclasses of a given reference class, without consideration of whether they can be defined in terms of selections by associated sequences that employ only admissible selective classes. Fetzer's conclusion would be correct if homogeneity were characterized on the basis of all kinds of selections, but we have known since von Mises's introduction of the notion of a place selection that certain restrictions must be imposed upon the kinds of selections employed in characterizing randomness. My aim in this chapter is to arrive at a suitable set of restrictions of this kind. The job, as we have seen, is not easy.

SOME PHILOSOPHICAL REFLECTIONS

Our definition of objective homogeneity involves considerations of physical randomness as well as mathematical randomness. In definition 3, it will be recalled, we stipulated that the attribute C, which is to be used to define a selection in A, must occur in D in a mathematically random way. There are certain reasons that strongly suggest it would be desirable, if possible, to drop this reference to mathematical randomness, and to define homogeneity solely in terms of physical relationships. Suppose, for example, that an apparently symmetrical coin is tossed in what appears to be the standard manner, but that the result is a perfectly alternating sequence of heads on every odd toss and tails on every even toss. This lack of *mathematical* randomness would surely be taken as strong evidence that there is something physically inhomogeneous about the sequence of tosses, and we would presumably undertake careful investigation of the physical mechanisms involved. We might find, for example, that the coin is placed in the flipping mechanism with alternate faces uppermost before each flip, and that when this pattern is altered the sequence of results is correspondingly broken up. In such cases, we might say, the lack of mathematical randomness in the sequence of results does not establish, *by definition*, that the class of tosses is inhomogeneous; instead, it leads us to search out a physical inhomogeneity—which can, of course, be described in terms of an associated sequence—and it is the lack of invariance with respect to such a selection by associated sequence that definitively establishes the lack of homogeneity. To incorporate a condition of mathematical randomness into the definition of physical homogeneity is, according to this line of thought, a confusion of empirical evidence with defining characteristics. Our basic concern, in raising the question of homogeneity in the first place, was to try to achieve clarity regarding such matters as indeterminism in the physical world and irreducibly statistical laws of nature. Mathematical randomness seems to be something of a red herring in this context.

A further argument reinforces the same point.[9] In all of the foregoing discussions, it has been necessary officially to take the classes A and D as ordered sequences of events. In many cases, the ordering may have seemed quite arbitrary. Such ordering was unavoidable, however, because definitions of randomness apply to infinite sequences and limiting frequencies within them. In infinite sequences, it is necessary to fix the order in which the events occur, for if the order is altered, the limiting frequency of a given attribute may change as a result. For finite classes, in contrast, the overall relative frequency of an attribute within a class is independent of the order in which the members are considered. This rather arbitrary aspect of the limiting frequency concept of probability has long been recognized. If we were to omit from our definition of an admissible selective class (definition 3) the requirement of mathematical randomness (clause (2)),

---

[9] I am indebted to Henry Krips for first pointing out this advantage of dropping the condition of mathematical randomness.

the result would be an identification of admissible selective classes with objectively codefined classes. This would, by implication, remove any requirement of mathematical randomness from the definition of objectively homogeneous reference classes. We could then omit consideration of the internal order of our reference classes. This would enable us to apply our concepts directly to finite classes, without always appealing to the fiction of hypothetical infinite sequences. In speaking about associated sequences, it is, of course, essential to preserve the appropriate coordination between the individual xs and the corresponding ys, but the internal orders of A and D need not be subject to any further mathematical constraints.

Freeing the concept of objective homogeneity from formulations involving infinite sequences with fixed internal order certainly appears to be a desirable goal. I would readily agree to do so—by omitting clause (2) of definition 3—were it not for the difficulty posed by example 4 (the superstitious crapshooter). Of all the problem-cases we have considered, it is the only one that seems to involve in any *essential* way the ordering of A and D. This example, it will be recalled, was an adaptation to associated sequences of a difficulty raised long ago in connection with von Mises's original definition of the collective. Perhaps there is a way of dealing with this problem without getting involved with mathematical randomness.

In his original presentation, von Mises distinguished two types of place selections. The first type, which I shall call "purely ordinal," qualifies an element x_i as belonging or not belonging to the selection S strictly on the basis of the value of its subscript i, regardless of the outcomes of earlier elements of the sequence. Selections of the other type take account of the attributes of preceding members of the sequence. Inasmuch as our definition of "associated sequence" (definition 1) leaves open the possibility that the y_i coordinated with an element x_i may be a preceding element of A— that is, y_i may be identical to x_j, where $j < i$—place selections of the second type mentioned by von Mises can be regarded as selections by associated sequences. Our definition of objective homogeneity excludes reference classes that can be relevantly partitioned by such selections even if there is no appeal to considerations of mathematical randomness. The use of mathematical randomness in defining admissible selective classes is needed solely to deal with reference classes that can be relevantly partitioned *only* by purely ordinal selections. Any reference class that can be relevantly partitioned by any other kind of selection is already disqualified from the status of objective homogeneity.

Let us take another look at example 4. The selective class C invoked in that case was defined in terms of the properties of some real number r. If we were to program the computer in our computer/detector apparatus to ascertain whether a particular y_i belongs to C, we would program it to calculate the binary expansion of r to the ith place, and to qualify y_i as a member of C if and only if the ith digit is 1. This program would not make any use of any physical property of y_i that could be furnished by the detector component of the apparatus. Examination of the program would show that for every value of i, y_i has the same status with regard to membership in C regardless of whether y_i possesses the property F to which the detector responds, and regardless of the pattern of F and $\bar{F}$ exhibited by preceding members of the associated sequence D. For purposes of this example, we are assuming that there does not exist any *physical* property F that can be detected by a physically possible detecting apparatus that would yield the same selective class C as was numerically described in terms of the binary expansion of r.

There is a strong temptation, at this point, simply to declare that objective homogeneity does not depend upon selections of the purely ordinal variety—that objectively homogeneous reference classes need not be invariant with respect to purely ordinal selections. If this were done, it would sever the connection between mathematical randomness and physical randomness (in the form of objective homogeneity). In that case, we could simplify the series of definitions leading up to definition 5. The requirement of infinitude could then be removed from the definition of an associated sequence (definition 1, p. 61) and the definition could be modified accordingly:

DEFINITION 1': *Let A be a reference class consisting of a sequence of events x_1, x_2, ... Any other sequence D consisting of events y_1, y_2, . . . will be called an* associated sequence *if each event y_i occurs in the absolute past (past light cone) of the corresponding event x_i.*

If A and D are finite, their ordering is arbitrary as long as the coordination between the two sequences remains fixed. The concept of an objectively codefined class (definition 2, p. 68) would remain unchanged. When clause (2) of definition 3 is dropped, admissible selective classes are identified with objectively codefined classes, making definition 3 otiose. The admissible selective classes used to define a selection in the reference class A (definition 4, pp. 69–70) would be codefined classes, allowing us to reformulate the definition:

DEFINITION 4': *Let the class y_1, y_2, . . . constitute an associated sequence D with respect to the reference class A (as detailed in definition 1'). Then, a selection by an associated sequence S is any selection within A defined by the rule*
$x_i \in S$ iff $y_i \in C$,
where C is an objectively codefined class (as detailed in definition 2).

Definition 5 remains unchanged: a reference class *A* is objectively homogeneous with respect to *B* if the probability of *B* in *A* is invariant with respect to all selections by associated sequences, where we understand, of course, that such selections are of the kind specified by definition 4′ instead of definition 4.

The main problem with the foregoing program is to find some justification other than wishful thinking for adopting it. The most plausible approach, it seems to me, is to introduce a *physical assumption* to the effect that while God may play dice with the world, He does not work numerology upon it. His dice generate random sequences; purely numerical considerations do not generate *physical regularities*.[10] It is logically possible, of course, that God chooses some computable number *r*—one that may be terribly complicated to calculate—to govern a given physical sequence *A*, but that there is no physical selection by an associated sequence that will yield a relevant partition of that reference class. Recalling example 4, it may be that a computable number *r* governs the outcomes of the tosses of the dice, giving rise to a physical regularity among these outcomes, but that no antecedent physical regularity reveals it. I do not believe, however, that this is how the world works. If the sequence is nonrandom in the mathematical sense, I am confident that there is some *physical* regularity that accounts for this regularity. Let us consider a more serious case.

> EXAMPLE 8: Suppose that a succession of electrons impinge, one after another, upon a potential barrier, and that their kinetic energy is just sufficient to give each of them a 50-50 chance of tunneling through. We assign that digit 1 to the passage of an electron through the barrier, and the digit 0 to the reflection of an electron from the barrier. In this case—unlike example 5, where we may eventually run out of radioactive atoms in our sample—we can plausibly treat the sequence as potentially infinite, for electrons can be recycled through the experiment as often as we wish.

There are two genuine possibilities, I believe. (1) Perhaps the standard quantum theory is correct and there are no 'hidden variables' to determine which electrons tunnel through and which are reflected. In this case, the sequence of 1s and 0s is random, and the reference class of impinging electrons is homogeneous with respect to tunneling. (2) The standard quantum theory is mistaken and there are as yet unknown *physical parameters* that determine passage and reflection. In this case, the sequence of 1s and 0s may or may not be random (depending upon initial conditions), but the

---

[10] Peter Railton (1978, pp. 218–219) seems to be making a somewhat similar assumption.

class of impinging electrons is not homogeneous, for it is not invariant under selections by associated sequences. A third alternative, which I cannot really take seriously, is the notion that the reference class is physically homogeneous—that is, invariant under selections generated by co-defined classes—but that the sequence of 1s and 0s is not mathematically random. It is my firm conviction that the real number corresponding to the sequence—and, of course, there is one—is not computable. It should be recalled that there are uncountably many real numbers between zero and one, but only a denumerable infinity of computable numbers. God does not pick a computable real number, and without giving any physical clue to what it is, use it to determine which electrons tunnel through and which are reflected. This is the assumption I propose to adopt.

The foregoing considerations presuppose, of course, that our sequences of events are potentially infinite. They will not work for finite classes, for every finite sequence of 1s and 0s is computable. There is, however, one way of extending the idea to finite sequences. The notion of *computational complexity* has been defined to deal with just such problems. We can say, roughly speaking, that a sequence has a high degree of computational complexity if the shortest program that will generate the sequence is of about the same length as the sequence itself. If, however, there is a very short program that will generate the sequence, it is computationally simple. Computational complexity is an excellent analogue of randomness for finite sequences. Applying this notion to example 8, we can say that any long finite sequence generated by the behavior of the electrons will be computationally complex.

It is my personal view that the foregoing assumption is acceptable, and that we can arrive at a satisfactory concept of objective homogeneity of reference classes without bringing in mathematical randomness. Thus I would accept the explication that results from definitions 1', 2, 4', and 5. This allows us to separate physical randomness from mathematical randomness, and it allows us to deal straightforwardly with finite classes of physical events. Whether others will be equally willing to embrace this simplification, I do not know. For those who are not, the explication given by definitions 1–5 is available.

Whatever attitude we finally adopt regarding the relationship between mathematical randomness and objective homogeneity, it seems clear that the concept of selection by an associated sequence furnishes the main consideration. It is easy to think up commonplace examples of reference classes that fail to be genuinely homogeneous because of the possibility of making selections on the basis of some sort of associated sequence. The class of tosses of a coin was seen to be inhomogeneous if we take note of the possibility of using the state of the coin immediately prior to its landing for purposes of making a selection. The same kind of consideration applies to the roulette wheel; as the wheel slows down and the ball is just about to fall, it is possible to predict the outcome with some reliability. Most of us believe it is possible in principle, though perhaps technically impossible at present, to predict with high reliability which victims of latent untreated syphilis will develop paresis and which will not.

Are there, in fact, any nontrivial, objectively homogeneous reference classes? No one knows for sure, but there seems to be a strong possibility that cases similar to examples 5 and 8, which exist in the quantum domain, embody objective homogeneity. Given a collection of heavy atoms, we can, in principle, sort them into different elements, and then into different isotopes of these elements. Some of the isotopes are stable; others have half-lives ranging from thousands of millions of years down to tiny fractions of a second. Thus the original collection is highly inhomogeneous with respect to the occurrence of spontaneous radioactive decay within a specified time span. If, however, we select only those atoms that belong to one isotope—say U^{238}—there is, to the best of our knowledge, no further partition that is relevant to spontaneous decay. There are, moreover, theoretical reasons for supposing that no as-yet-undiscovered property possessed by the nuclei prior to decay is statistically relevant to spontaneous decay. If this is true, the physical world does, indeed, contain objectively homogeneous reference classes. It is my hope that we can at least *assign a reasonable meaning to such a statement*, whether it happens to be true or false.

If we have succeeded in characterizing objective homogeneity, then I think we can make sense of certain other fundamental concepts. If, as suggested in the preceding paragraph, there are objectively homogeneous reference classes of the nontrivial variety, then the physical world is objectively indeterministic. If the world is indeterministic, then some physical laws are irreducibly statistical. Let us assume for the moment—admittedly, no small assumption—that we know what it means to say that a given statement is a statistical law. If we employ reasonable classes A and B, then presumably the statement "$P(B|A) = p$" would qualify as a statistical law, though not necessarily an irreducible one. I see no reason to deny the possibility of bona fide statistical laws that are not irreducible, such as we would have in classical statistical mechanics even if all molecular motions conformed to strict deterministic laws. If, however, the reference class A in the foregoing statistical law is objectively homogeneous with respect to the attribute B, and if some As are B and some are not, then the foregoing law would qualify as irreducibly statistical.[11] If we are going to

---

[11] Some of the difficulties encountered by J. L. Mackie in his discussion of statistical laws

make sense of scientific explanation in the context of twentieth-century science, we will need serviceable concepts of *objective homogeneity* and *irreducibly statistical law*. I hope this chapter comes somewhere near to furnishing them. The concept of physical randomness with which we have been dealing seems indispensable to our scientific understanding of the world.

(1974, chap. 9) seem to point clearly to the need for a viable concept of objective homogeneity. I suspect that Railton (1978), at least implicitly, needs the same concept; see footnote 4 of this chapter.

4 | The Three Conceptions Revisited

IN THE first chapter, after considering some of Laplace's remarks pertaining to scientific explanation, I extracted three general conceptions that seemed to be present in his thought on the subject. These three conceptions, which are shared by many other historical and contemporary writers, were examined in the context of classical determinism. In that context they seemed hardly distinguishable from one another. While noting that fact, I claimed that we would find sharp divergences among them when we made the transition from the deterministic context, in which explanations rely only upon universal laws, to the *possibly indeterministic* context of twentieth-century science, in which it may be necessary to invoke statistical laws in constructing scientific explanations. Now that we have had the opportunity to take a fairly close look at statistical explanation, it is time to examine the three general conceptions in this more up-to-date context.

THE EPISTEMIC CONCEPTION

When the three basic conceptions were introduced in the first chapter, I pointed out that the approach that holds scientific explanations to be arguments has been the dominant view for several decades, and that it currently enjoys the status of the 'received doctrine.' Indeed, in "A Third Dogma of Empiricism" (Salmon, 1977a), it played the title role. The notion that explanations are arguments is clearly an epistemic conception, and it was presented as such in chapter 1. Let us refer to it, more precisely, as the *inferential version* of the epistemic conception. In the present chapter, where the three basic conceptions will be discussed in greater detail, we shall distinguish two other versions of the epistemic approach—the *information-theoretic version* and the *erotetic version*—which have arrived much more recently upon the scene. Because of its enormous influence, the inferential version has first claim upon our attention.

The Inferential Version

The aim of scientific explanation, according to the inferential version of the epistemic conception, is to provide *nomic expectability* for the event-to-be-explained. In the deterministic context, this meant that the occurrence of the explanandum-event could be shown to be *deductively certain* relative to the explanatory facts. When we leave this deterministic context, it is natural to understand nomic expectability in terms of a *high inductive probability* of the explanandum relative to the explanans. Deductive certainty goes over into high inductive probability. The major problem that arises in this transition, it seems to me, is the problem of explaining low-probability events.

When this question of explaining low-probability events was first introduced in chapter 2, I mentioned paresis and mushroom poisoning as two examples. As I noted in chapter 3, a natural response to examples of that sort might be to regard them as partial explanations, to be completed when our medical knowledge of such phenomena has been significantly expanded. Indeed, it might plausibly be maintained that even if some phenomena occur in an irreducibly statistical manner, these particular examples are unlikely to be cases of that type. They will probably be amenable, at some future time, to deductive-nomological explanations in terms of universal laws. In this connection, it might be noted that Hempel's example of the strep infection, which he introduced as a basic instance of statistical explanation, has subsequently enjoyed a change of status, for it is now possible to provide a full deductive account of the penicillin resistance of microorganisms.[1] I shall not argue that paresis and mushroom poisoning are ineluctable cases of low-probability events.

There is, however, no difficulty in producing examples that exhibit similar low-probability values, and that are much less plausible candidates for treatment as incomplete explanations. Suppose, for instance, that an electron, which has a given amount of kinetic energy, approaches a potential barrier of such a magnitude that the electron has a 3/4 chance of being transmitted through the barrier and a 1/4 chance of being reflected back. Given an electron that does return in the direction from which it came, I think it is perfectly reasonable to explain its return as a result of an encounter with the potential barrier. Since, as I understand it, about 1/4 of all victims of latent untreated syphilis develop paresis, we can count the reflection of the electron as a possible case of an irreducibly statistical phenomenon exhibiting about the same low probability. If only one person in a hundred is adversely affected by consuming a particular type of mushroom, we can easily construct a parallel example by simply altering the height of the potential barrier in the foregoing example. Further adjustments will obviously furnish probability values of any magnitude we choose. As I have often mentioned, when an alpha particle in the nucleus of a U^{238}

---

[1] The chemistry of bacterial resistance to penicillin seems now to be understood (S. N. Cohen, 1975).

atom approaches the potential barrier that constitutes the wall of the nucleus, it has a probability of about 10^{-38} of tunneling through, but when such a spontaneous radioactive decay occurs, the explanation is that a low-probability event has occurred. According to the standard statistical interpretation of quantum mechanics, all such transmissions and reflections by particles encountering potential barriers are irreducibly statistical events.

There is, of course, an obvious response that can be made by the defender of the high-probability requirement—namely, that under the foregoing circumstances no explanations have been given. If further relevant factors could be found, in relation to which these events could be rendered highly probable, then they could be explained. Unless we can find additional relevant factors, such events are simply not amenable to explanation. This kind of response leads, in my opinion, to certain undesirable consequences, which can be illustrated by a Mendelian genetic experiment on the color of pea blossoms (Hempel, 1965, pp. 391–392). In one population, for example, 3/4 of the blossoms are red and 1/4 are white. Suppose I point to a particular red blossom and ask, ''Why does that one have the color it has?'' We can answer that it comes from this population, and therefore it has a fairly high probability of being red. Suppose, however, that I point to a particular white blossom and ask why it has that color. The answer must be that there simply is no explanation; the event is inexplicable because it is improbable. This response seems completely unacceptable. I find it hard to believe that we do not understand the occurrence of white blossoms in this population just as adequately as we understand the occurrence of red.

This point, which has fundamental importance, can be generalized. If some events are probable, but not necessary, then other events are improbable. Suppose, for instance, that a coin is heavily biased for heads, so that the probability of heads is 0.9, while the probability for tails is 0.1. If we assume that 0.9 is sufficiently large to qualify as a high probability for purposes of statistical explanation, then we can construct an I-S explanation of any head outcome. However, tails also occurs, though with a much smaller frequency. If we accept the high-probability requirement, then we must say that the tail outcomes cannot be explained. There is a strange lack of parity here. As I remarked in connection with the pea blossom experiment, it appears that we understand the improbable outcome just as well as we understand the highly probable outcome. It strikes me as a rather peculiar prejudice which says that we can explain phenomena that occur frequently, but not those that occur infrequently. This point is argued persuasively by Richard Jeffrey (1969), who takes a genetic example as one of his main illustrations; it is reargued by Peter Railton (1978), who provides detailed treatment of such quantum mechanical phenomena as barrier penetration and spontaneous radioactive decay. These considerations lead both Jeffrey and Railton to reject Hempel's I-S model of inductive explanation and to deny the basic thesis that statistical or probabilistic explanations are arguments. Since, moreover, both of these authors rightly emphasize the indispensable role of physical processes and mechanisms in scientific explanation, they can, I believe, be appropriately included among the proponents of the ontic conception.

Imposition of the high-probability requirement upon explanations produces a serious malady, which Henry Kyburg (1970) has dubbed "conjunctivitis"; Hempel was fully aware of this problem and discussed it explicitly (1965, pp. 410–412). Because of the basic multiplicative rule of the probability calculus, the joint occurrence of two events is normally less probable than either event individually. This is illustrated by the previously mentioned biased coin. Suppose we decide that, for explanatory purposes, a probability of 0.9 or greater will qualify as high, while smaller probabilities will not. If the coin is tossed twice, with the result of head on each toss, we could claim to have an I-S explanation of each result individually. However, since the probability of tossing two heads in a row is $0.9 \times 0.9 = 0.81$, the joint occurrence would be unexplainable. The moral to be drawn from examples of this kind is, it seems to me, that there is no reasonable or nonarbitrary way of answering the question "How high is high enough?"

If conjunctions are the enemies of high probabilities, disjunctions are their faithful allies. If a fair coin is tossed ten times, there is a probability of 1/1,024 that it will come up heads on all ten tosses. This sequence of events constitutes a (complex) low-probability event, and as such, it is unexplainable under any sort of high-probability requirement. Even if the outcome is five heads and five tails, however, the probability of that sequence of results, *in the particular order in which they occurred*, is also 1/1,024. It, too, is a low-probability event, and as such, it is unexplainable. If, however, we consider the probability of five heads and five tails *regardless of order*, we are considering the disjunction of all of the 252 distinct orders in which that outcome can occur. Even this extensive disjunction has a probability of about 0.246; hence, even it fails to qualify as a high-probability event. If, however, we consider the probability of getting *almost* 1/2 heads in ten tosses—that is, four or five or six heads— this disjunction is ample enough to have a probability somewhat greater than 0.5 (approximately 0.656). The general conclusion would seem to be that, for even moderately complex events, every specific outcome has a low probability, and is consequently incapable of being explained.[2] The

---

[2] It seems to me that Baruch Brody (1975, p. 71) missed this point when he wrote: "It

only way to achieve high probabilities is to erase the specific character of the complex event by disjunctive dilution.

Richard Jeffrey (1969) and James Greeno (1970) have both argued, quite correctly, I believe, that the degree of probability assigned to an occurrence in virtue of the explanatory facts is not the primary index of the value of the explanation. Suppose, for example, that two individuals, Sally Smith and John Jones, both commit suicide. Using our best psychological theories, and summoning all available relevant information about both persons—such as age, sex, race, state of health, marital status, religion, and so forth—we find that there is a low probability that Sally Smith would commit suicide, whereas there is a high probability that John Jones would do so. This does not mean that the explanation of Jones's suicide is better than that of Smith's, for exactly the same theories and relevant factors have to be taken into account in both.

According to the S-R approach outlined in chapter 2, the statistical basis of scientific explanation consists, not in an argument, but in an assemblage of relevant considerations. High probability is not the desideratum; rather the amount of relevant information is what counts. According to this approach, the S-R basis of statistical explanation consists of a probability distribution over a homogeneous partition of an initial reference class. A homogeneous partition, it will be recalled, is one that does not admit of further relevant subdivision. The subclasses in the partition must also be maximal—that is, the partition must not involve any irrelevant subdivisions. The goodness, or epistemic value, of such an explanation is measured— as we shall see more fully when we discuss the information-theoretic version of the epistemic conception—by the gain in information provided by the probability distribution over the explanandum-partition relative to

should be noted that there are some cases of statistical explanation where the explanans does provide a high enough degree of probability for the explanandum, so Hempel's requirements laid down in his inductive-statistical model are satisfied, but does not differentiate between the explanandum and some of its alternatives. Thus, one can explain, even according to Hempel, the die coming upon one in 164 of 996 throws by reference to the fact that it was a fair die tossed in an unbiased fashion; such a die has, after all, a reasonably high probability of coming upon one in 164 out of 496 [sic] throws. But the same explanans would also explain its coming upon one in 168 out of 996 throws. So it doesn't even follow from the fact that an explanation meets all of Hempel's requirements for statistical explanations that it meets the requirement that an explanans must differentiate between the explanandum and its alternatives.'' The fact is that a fair die, tossed 996 times in an unbiased fashion, has a probability of about 0.0336 of showing side one in 164 throws. By no stretch of the imagination can this be taken as a case in which Hempel's high-probability requirement is satisfied. The probability that in the same number of throws with the same die the side one will show 168 times is nearly the same, about 0.0333. As Hempel has observed, if two events are incompatible, they cannot both have probabilities that are over 0.5 on the same consistent body of evidence—a minimal value, I should think, for any probability to qualify as high.

the explanans-partition.[3] If one and the same probability distribution over given partitions of a reference class provides the explanations of two separate events—one with a high probability and one with a low probability—the two explanations are equally valuable.

In 1977, Hempel published a German translation of his major essay, "Aspects of Scientific Explanation," and into this translation he inserted, at the end of the section on statistical explanation, a postscript, "Nachwort 1976," dealing with certain issues that had arisen since the original English version of "Aspects" was published in 1965. No English translation of this postscript has as yet been published. The 1976 postscript contains three major points that have direct bearing upon the issues we are discussing.

(1) As mentioned in chapter 3, Hempel modifies his doctrine of *essential* epistemic relativization of inductive-statistical explanation, allowing that in rare cases we may be able to find objectively homogeneous reference classes. This means that it makes sense to talk about objective homogeneity, and to claim (truly or falsely) that we have an I-S explanation simpliciter, not relativized to any particular knowledge situation.

(2) Hempel relinquishes the high-probability requirement, allowing that we may have I-S explanations of improbable events. It may seem hard to square the abandonment of the high-probability requirement with the epistemic conception of scientific explanation, which regards I-S explanations as inductive arguments, but as we shall see in connection with the next item, he offers what is to me a rather surprising clarification of the concept of inductive inference, which makes it possible to reconcile these two ideas.[4]

(3) Hempel has often reiterated the claim that I-S explanations are arguments, and he has frequently furnished schemata, similar to those I presented in chapter 2, which purport to exhibit the structure of I-S explanations. Such schemata contain premises and conclusions separated from one another by a double line. One would naturally suppose, I should think, that such arguments would be governed by some sort of rule of acceptance, according to which one could—at least tentatively—accept the conclusion as an item of knowledge if one is sufficiently confident of the truth of the premises and if the degree of inductive probability with which the conclusion is supported by the premises is sufficiently high. As I explain in detail in Salmon (1977b), Hempel seems never to have given any clear indication, in any of his writings on scientific explanation prior to 1976,

---

[3] For a discussion of measures of the values of explanations, one based upon degree of homogeneity and one based upon amount of information transmitted, see (Salmon et al., 1971, pp. 10–16, 51–53, 89–104).

[4] In this discussion, I am using the terms "inference" and "argument" interchangeably.

that he construed the notion of an inductive argument in any other way. Nevertheless, in the 1976 postscript, he states that he never intended this straightforward interpretation.

The issue arises in connection with a technical feature of Carnap's system of inductive logic, which does not admit inductive acceptance rules. Instead, according to Carnap, inductive logic furnishes degree of confirmation statements that assign a numerical value to the inductive probability of a given hypothesis on the basis of given evidence. However, one is never permitted to detach the hypothesis and assert it independently—no matter how high the degree of inductive probability and no matter how certain the evidence. Rather, if the evidence embodies all available relevant evidence, then the degree of confirmation of the hypothesis upon that evidence furnishes a fair betting quotient for wagering upon the truth of the hypothesis. One important feature of Carnap's system is that high probabilities are no better than low probabilities; whether the degree of confirmation is high, middling, or low, it provides an index to what constitutes a rational bet. Since there is never any question of detaching an inductive conclusion, we never have to ask whether the inductive probability is high enough to justify asserting the hypothesis as an independent item of knowledge.

If, as he says, Hempel wants to construe "inductive argument" along these Carnapian lines, he can give up the high-probability requirement with impunity. Instead of saying that an I-S explanation is an argument to the effect that the event-to-be-explained was to be expected in view of the explanatory facts, he can say that such an 'argument' tells us *the degree to which* that event was to be expected. This approach strikes me as a distinct improvement over the high-probability requirement, but it does have a serious pitfall. On Carnap's system of inductive logic, one can have degree of confirmation statements that assign inductive probabilities to hypotheses about particular events relative to evidence exclusively about other particular events. No universal or statistical generalizations are involved, nor could they be involved. Evidence statements must be ones we are prepared to accept, and if inductive logic contains no rules of acceptance, then *general statements* could *never* be *accepted* into our body of evidence. Consequently, if one were to adopt an inductive logic that has this feature, it would be necessary to give up the 'covering law' conception of scientific explanation, for there would be no reason—and indeed no way—to invoke laws in such explanations (see Salmon, 1968). This fact, which seems to be an unavoidable consequence of adopting an inductive logic that precludes rules of acceptance, has not, as far as I know, been squarely faced by Hempel. To me, this seems too high a price to pay to retain the thesis that scientific explanations of the statistical sort are arguments.

Because the doctrine that scientific explanations are arguments has held such a strong grip on twentieth-century philosophers of science, perhaps a little more should be said about its appeal. The fact is that many scientific explanations seem clearly to take the form of arguments. Although proponents of this view often cite examples in which some particular fact is explained—for example, the yellow color of this Bunsen flame at this time—it receives its strongest support from cases in which one generalization is explained by subsumption under another broader generalization. Some of the most beautiful explanations furnished by modern (post-Copernican) physics exemplify this point: for example, Newton's explanation of the approximate correctness of Kepler's laws of planetary motion in terms of his law of universal gravitation and his three laws of motion, Maxwell's explanation of optical phenomena in terms of his theory of electromagnetic radiation, and the relativistic explanation of the magnetic force felt by a moving charge in terms of the Lorentz contraction of the electric field through which it moves. In each case, a general proposition about a class of phenomena is established by deducing it from more general and more fundamental laws. Any philosophical theory of scientific explanation that could not accommodate such paradigms would be patently inadequate.

The main difficulty in this approach, I think, is that it involves a misconstruction of the nature of subsumption. One obvious way in which subsumption can be *exhibited* is in terms of an argument—to deduce "Socrates is mortal" from "All men are mortal" and "Socrates is a man" surely shows that the mortality of Socrates is subsumed under the mortality of humanity in general. Deductive relations, however, hold between statements or propositions, but the *fact* of Socrates' mortality is a nonlinguistic matter, as is the general law about the mortality of all humans. While we may formulate statements to articulate these facts, the objective relation of subsumption can be taken as holding between nonlinguistic facts, quite apart from any linguistic expression of these facts and quite apart from anyone's knowledge of them. When we are discussing scientific explanation, it seems advisable to think of the event-to-be-explained as an instance of a regularity in nature (a pattern), rather than to focus attention upon a logical relation among statements describing these particular and general facts. We have, perhaps, gone too far in our efforts to treat all philosophical problems in the "formal mode" rather than the "material mode."[5]

---

[5] I am in strong agreement with van Fraassen on the undesirability of excessive emphasis upon linguistic issues. "The main lesson of twentieth-century philosophy of science may well be this: no concept which is essentially language-dependent has any philosophical importance at all" (1980, p. 56).

In chapter 2 I discussed the statistical-relevance (S-R) model of scientific explanation.[6] Providing an explanation of this type involves summoning a set of statistical laws that express objective relations of statistical relevance. As we have seen, a *particular explanandum-event* can be subsumed under such general regularities without involving in any way the construction of arguments. If, in some cases, the regularities are related to the particular facts in ways that make it possible to construct deductive or inductive arguments, that is entirely beside the point as far as explanation of those facts is concerned.

Similar remarks can be made about the subsumption of a *general regularity* under another general regularity. According to classical physics, all material bodies in the universe are subject to Newton's law of universal gravitation and his three laws of motion. Now, if among all of these material bodies there should be one or more instances in which a relatively light body is moving in an orbit around another much more massive body (far enough away from other bodies of large mass to be reasonably unperturbed by them), then the motion will be of the kind described by Kepler's laws. While it is true that Kepler's laws can be deduced from Newton's laws, it is also correct to say that the physical regularity exhibited by these specially situated bodies is part of the general regularity exhibited by all bodies of any sort that possess gravitational mass. The more restricted regularity (which is not a statement, but a physical fact) is part of the more general pattern (which is also a physical fact). Indeed, whenever the physical part-whole relation obtains, that is also a physical fact that is entirely independent of the behavior of language-users and of the epistemic states of scientists. Thus there is no obstacle to our acceptance of the thesis that scientific explanation involves subsumption under laws and our rejection of the view that explanations are arguments. It is the *physical* subsumptive relation, not the inferential relations of deductive or inductive logic, which is exhibited by our beautiful scientific paradigms. The supposition that relations of subsumption *must* be interpreted in terms of logical argument forms is, I believe, one of the most unfortunate errors in modern philosophy of science.

There are, it seems to me, two questions that should be posed to those who still feel inclined to suppose that being an argument is an essential feature of every acceptable scientific explanation.

QUESTION 1: *Why are irrelevancies harmless to arguments but fatal to explanations?*

---

[6] Even though I now reject the S-R model of scientific explanation, this point about the nature of subsumption still applies.

In deductive logic, irrelevant premises are pointless, but they do not undermine the validity of the argument. Even in the relevance logic of Anderson and Belnap, p & $q \vdash p$ is a valid schema. If one were to offer this argument,

All men are mortal.
Socrates is a man.
Xantippe is a woman.

Socrates is mortal.

it would seem strange, and perhaps mildly amusing, but its logical status would not be impaired by the presence of the third premise. There are more serious examples. When it was discovered that the axioms of the propositional calculus in *Principia Mathematica* were not all mutually independent, there was no thought that the logical system was thereby vitiated. Nor is the validity of propositions 1–26 of Book 1 of Euclid's *Elements* called into question as a result of the fact that they all follow from the first four postulates alone, without invoking the famous fifth (parallel) postulate. This fact, which has important bearing upon the relationship between Euclidean and non-Euclidean geometries, represents a virtue rather than a fault in Euclid's deductive system.

When we turn to deductive explanations, however, the situation is radically different. The rooster who 'explains' the rising of the sun on the basis of his regular crowing is guilty of more than a minor logical inelegancy. So also is the person who 'explains' the dissolving of a piece of sugar by citing the fact that the liquid in which it dissolved is *holy* water. So also is the man who 'explains' *his* failure to become pregnant by noting that he has faithfully consumed his wife's birth control pills.[7]

The same lack of parity exists between inductive arguments and explanations. In inductive logic, there is a well-known requirement of total evidence (Carnap, 1950, sec. 45B). This requirement demands the inclusion of all relevant evidence. Since irrelevant 'evidence' has, by definition, no effect upon the probability of the hypothesis, inclusion of irrelevant premises in an inductive argument can have no bearing upon the degree of strength with which the conclusion is supported by the premises.[8] If facts of unknown relevance turn up—for example, a button found by a detective at the scene of a crime, which may or may not have any bearing

[7] These examples, and many others like them, can be schematized so as to fulfill all of Hempel's requirements for deductive-nomological explanation. This is shown in detail in (Salmon et al., 1971, pp. 33–35).

[8] If "$c(h,e)$" designates the inductive probability or degree of confirmation of hypothesis h on evidence e, then i is irrelevant to h in the presence of e if and only if $c(h,e.i) = c(h,e)$.

upon the identity of the criminal—inductive sagacity demands that they be mentioned in the premises. No harm can come from including them if they are irrelevant, but considerable mischief can result if they are relevant and not taken into account.

When we turn our attention from inductive arguments to inductive explanations, the situation changes drastically. If the consumption of massive doses of vitamin C is statistically irrelevant to immunity to the common cold, then 'explaining' freedom from colds on the basis of the use of that medication is worse than useless. So also would be the 'explanation' of psychological improvement on the basis of psychotherapy if the spontaneous remission rate for neurotic symptoms were equal to the percentage of 'cures' experienced by those who undergo the particular type of treatment.[9]

Hempel recognized from the beginning the need for some sort of requirement of total evidence for inductive-statistical (I-S) explanation; it took the form of the *requirement of maximal specificity* (Hempel, 1965, pp. 394–403). This requirement stipulates that the reference class to which an individual is referred in a statistical explanation be narrow enough to preclude, in the given knowledge situation, further relevant subdivision. It does not, however, prohibit irrelevant restriction. I have therefore suggested that this requirement be amended as the *requirement of the maximal class of maximal specificity* (Salmon et al., 1971, pp. 47–51). This requirement demands that the reference class be determined by taking into account all relevant considerations, but that it not be irrelevantly partitioned. Hempel (1977, pp. 107–111) has rejected this suggested modification.

Inference, whether inductive or deductive, demands a requirement of total evidence—a requirement that *all* relevant evidence be mentioned in the premises. This requirement, which has substantive importance for inductive inferences, is automatically satisfied for deductive inferences. Explanation, in contrast, seems to demand a further requirement—namely, that *only* considerations relevant to the explanandum be contained in the explanans. This, it seems to me, constitutes a deep difference between explanations and arguments.

QUESTION 2: *Why should requirements of temporal asymmetry be imposed upon explanations (while arguments are not subject to the same constraints)?*

---

[9] Examples of this sort are discussed in (Salmon et al., 1971, pp. 33–35), where they are shown to conform to Hempel's requirements for inductive-statistical explanation.

A particular lunar eclipse can be predicted accurately, using the laws of motion and a suitable set of initial conditions holding prior to the eclipse; the same eclipse can equally well be retrodicted using posterior conditions and the same laws. It is intuitively clear that if explanations are arguments, then only the predictive argument can qualify as an explanation, and not the retrodictive one. The reason is obvious. We explain events on the basis of antecedent causes, not on the basis of subsequent effects (or other subsequent conditions).[10] A similar moral can be drawn from Sylvain Bromberger's flagpole example. Given the elevation of the sun in the sky, we can infer the length of the shadow from the height of the flagpole, but we can just as well infer the height of the flagpole from the length of the shadow. The presence of the flagpole explains the occurrence of the shadow; the occurrence of the shadow does not explain the presence of the flagpole. At first blush, we might be inclined to say that this is a case of coexistence; that is, the flagpole and the shadow exist simultaneously. Upon closer examination, however, we realize that a causal process is involved, and that the light from the sun must either pass or be blocked by the flagpole *before* it reaches the ground where the shadow is cast.

In a charming piece of philosophy-of-science-fiction, van Fraassen, who places considerable emphasis upon the pragmatic aspects of explanation, tries to show that, in a certain context, the length of the shadow can explain the height of the flagpole (1980, pp. 132–134). What his story demonstrates is that a prior intention of someone to have a shadow of a given length can explain the height of the flagpole. None of us would have doubted that claim.

There are, of course, instances in which inference enjoys a preferred temporal direction. As a holiday weekend approaches, we can predict with confidence that many people will be killed in highway accidents, and perhaps we can give a good estimate of the number. We cannot, with any degree of reliability, predict the exact number—much less the identity of each of the victims. By examining the newspapers of the following week, however, we can obtain an exact account of the number and identities of these victims as well as a great deal of information about the circumstances of their deaths. On a Fourth of July weekend in Arizona, for example, ten people were killed in a tragic head-on collision. On a straight stretch of road, with clear visibility and no other traffic present, a pickup truck crossed the center line and struck an oncoming car. Prediction of such an event,

---

[10] The issue of temporal asymmetry, including such examples as Bromberger's flagpole, is discussed at some length in (Salmon et al., 1971, pp. 71–76). We shall return to this problem in chapter 6.

even ten minutes before its occurrence, would have been out of the question. By techniques of dendrochronology (tree ring dating), for another example, relative annual rainfall in parts of Arizona is known for some eight thousand years into the past. By way of contrast, no one could provide even a reasonable guess about relative annual rainfall for a decade into the future.

Such examples show that the temporal asymmetry reflected by inferences is precisely *opposite* to that exhibited by explanations. We have many records, natural and humanly made, of events that have happened in the past; from these records we can make reliable inferences into the past. We do not have similar records of the future. Prognostication is far more difficult than retrodiction; it has no aid comparable to records. The reason is that it is often possible to infer the nature of a cause from a partial effect, but it is normally impossible to infer the nature of an effect from knowledge of a partial cause. Yet no one would be tempted to 'explain' the accidents of a holiday weekend on the basis of their being reported in the newspaper. The newspaper account might in some cases *report* an explanation of an accident, but it surely does not *constitute* an explanation of the accident. Similarly, no one would be tempted to try to 'explain' the rainfall of past millennia on the basis of rings in bristlecone pine trees. If it were indeed true that being an argument is an essential characteristic of scientific explanations, how are we to account for the total disparity of temporal asymmetry in explanations and in arguments? This is a fundamental question for supporters of the inferential view of explanation.

If one rejects the inferential view of scientific explanation, it seems to me that straightforward answers can be given to the foregoing questions. On the statistical-relevance approach, the S-R basis of an explanation is an assemblage of factors that are statistically relevant to the occurrence of the event-to-be-explained. To offer an item as relevant when it is, in fact, irrelevant is clearly inadmissible. We thus have an immediate answer to our first question, "Why are irrelevancies harmless to arguments but fatal to explanations?"

When we come to the second question, regarding temporal asymmetry, we cannot avoid raising the issue of causation. In the classic 1948 essay, Hempel and Oppenheim suggested that D-N explanations are causal explanations, but in subsequent years Hempel has backed away from this position, explicitly dissociating 'covering law' explanations from causal explanations.[11] The time has come, it seems to me, to put the "cause" back into "because." Consideration of the temporal asymmetry issue forces

---

[11] See (Hempel, 1965, p. 250) for the 1948 statement, but see note 6 (added in 1964) on the same page. The later view is elaborated more fully in (Hempel, 1965, pp. 347–354).

reconsideration of the role of causation in scientific explanation, and of the grounds for insisting that occurrences are to be explained in terms of antecedent causes rather than subsequent effects. Causality will be treated at length in chapters 5–7.

The Information-Theoretic Version

As Greeno shows elegantly in his (1970), there is one plausible way of implementing the epistemic conception of scientific explanation without getting into all of the difficulties that plague the inferential version. Instead of regarding the goal of a scientific explanation as the achievement of 'nomic expectability' on the basis of deductive or inductive arguments, one can look at scientific explanations as ways of increasing our information about phenomena of the sort we are trying to explain. We can borrow the technical concept of *information transmitted* from information theory, and then go on to evaluate explanations in terms of the amounts of information that they transmit. Since "information" is defined probabilistically in information theory, this idea is especially well suited for dealing with statistical explanations.

The basic concepts involved in this approach are quite straightforward. For purposes of concrete illustration, let us return to Greeno's example of juvenile delinquency. Our general reference class A might be the class of American teen-agers. This class can be partitioned twice: once into an explanandum-partition $\{M\} = \{M_1, \ldots, M_m\}$, and again into an explanans-partition $\{S\} = \{S_1, \ldots, S_s\}$. As the final letters of "explanandum" and "explanans," respectively, "M" and "S" serve a mnemonic function. In the concrete example, we could let $\{M\} = \{M_1, M_2, M_3\}$, where M_1 is the subclass of teen-agers with no criminal convictions, M_2 is the subclass with only minor criminal convictions, and M_3 is the subclass with at least one conviction for a major crime. In Greeno's example, Albert, the youth in question, stole an automobile, so he belongs to M_3. The partition $\{S\}$ would consist of a large number of subsets defined in terms of such factors as sex, religious background, place of residence, socioeconomic status of family, and so forth. Using standard information-theoretic concepts, we may define $H(M)$, the uncertainty of the explanandum-partition, and $H(S)$, the uncertainty of the explanans-partition, as follows:

$$H(M) = \sum_{i=1}^{m} - p_i \log p_i$$

$$H(S) = \sum_{j=1}^{s} - p_j \log p_j$$

where

$$p_i = P(M_i|A) \text{ and } p_j = P(S_j|A).$$

The uncertainty of the system of partitions is given by

$$H(S \times M) = \sum_{i=1}^{m} \sum_{j=1}^{s} - p_i p_{ij} \log p_i p_{ij}$$

where

$$p_{ij} = P(M_j|A.S_i).$$

For this system of partitions, we define the information transmitted as follows:

$$I_T = H(M) + H(S) - H(M \times S).$$

In (Salmon et al., 1971, pp. 13–15), I offered the following simple example to illustrate Greeno's method. Suppose that the population of Centerville, U.S.A., is equally divided between Democrats and Republicans. Let us partition the population in terms of political affiliation:

$\{M\} = \{D,R\}; M_1 = D \text{ and } M_2 = R$
$p_1 = P(M_1) = P(D) = \frac{1}{2}$
$p_2 = P(M_2) = P(R) = \frac{1}{2}$

This partition involves the greatest possible uncertainty for a partition into two subclasses, for knowing that people are residents of Centerville tells us nothing about their membership in one party or the other. In this case (using logarithms to the base 2),

$$H(M) = - (\frac{1}{2} \times -1) - (\frac{1}{2} \times -1) = 1,$$

where it is sometimes called, with enormous potentiality for confusion, the "information." The bifurcation into two equally probable subsets provides the unit of uncertainty (or information) known as the "bit." Notice that the uncertainty drops to zero when either p_1 or p_2 assumes the value one. If all residents of Centerville were Republicans, there would be no uncertainty whatever about their party affiliation.

Now suppose, however, that Centerville is split by railroad tracks that run north-south through the town, and that half of the residents live to the east and half to the west of the tracks. Here we have another partition:

$\{S\} = \{E,W\}; S_1 = E \text{ and } S_2 = W$
$p'_1 = P(S_1) = P(E) = \frac{1}{2}$
$p'_2 = P(S_2) = P(W) = \frac{1}{2}$

Again, the uncertainty is maximal:

$$H(S) = 1.$$

If the two partitions $\{M\}$ and $\{S\}$ were statistically independent of one another, then knowing a person's place of residence with respect to the tracks would convey no information regarding party affiliation. According to our sociological folklore, however, there is a right and a wrong side of the tracks in any town, and it might be reflected in people's political persuasions. Let us therefore assume that 3/4 of the people who live on the east side of the tracks are Democrats, while 3/4 of those on the west side are Republicans. Since we are assuming that equal numbers of people live on the two sides of the tracks, we have the following conditional probabilities:

$$
\begin{aligned}
P(D|E) &= P(M_1|S_1) &= p_{11} &= \tfrac{3}{4} \\
P(R|E) &= P(M_2|S_1) &= p_{12} &= \tfrac{1}{4} \\
P(D|W) &= P(M_1|S_2) &= p_{21} &= \tfrac{1}{4} \\
P(R|W) &= P(M_2|S_2) &= p_{22} &= \tfrac{3}{4}
\end{aligned}
$$

Then,

$$H(S \times M) = -2(\tfrac{3}{8} \log_2 \tfrac{3}{8} + \tfrac{1}{8} \log_2 \tfrac{1}{8}) \simeq 1.81;$$
$$I_T \simeq 0.19.$$

Greeno shows that the increase in information is maximal when all of the conditional probabilities are either zero or one. This corresponds to the situation in which deductive-nomological explanation is possible. If, however, the marginal probabilities $P(M_i)$ of the explanandum-partition are also either zero or one, that maximum represents no gain in information. This situation corresponds to the case in which D-N explanation becomes vacuous because of a failure of relevance—that is, the conditional probabilities equal the marginal probabilities—as in Kyburg's hexed salt example. If, however, the explanans-partition is statistically relevant to the explanandum-partition there will be a gain in information. These considerations show that there is a close resemblance between Greeno's information-theoretic approach and the statistical-relevance model of scientific explanation.

As Greeno explicitly noted, the discussion in his (1970) was confined to statistical explanations of observable phenomena in terms of empirical statistical laws. At that level, this information-theoretic approach to the analysis of scientific explanation has many virtues. One of its main virtues is that—unlike the inferential version—it makes use of statistical relevance relations instead of degrees of inductive support (degrees of confirmation).

As mentioned previously, this version of the epistemic conception gets into no trouble at all over explanations of low-probability events.

Greeno realized from the beginning that the foregoing theory of scientific explanation is, at best, incomplete. As Joseph Hanna (1978) pointed out, statistical relationships among observables have little, if any, explanatory force. I am in complete accord with this criticism; indeed, it is basically the very argument that has led me to reject the S-R model as an adequate account of scientific explanation, and to insist that the S-R basis needs to be supplemented with causal and theoretical considerations in order to be able to characterize genuine scientific explanations. It therefore seems to me, in effect, that Greeno's initial information-theoretic account (1970) had just about the same strengths and weaknesses as the S-R model.

In a symposium devoted to "Theoretical Entities in Statistical Explanation" (1971), Greeno attempted to overcome the foregoing difficulty by extending his treatment to explanations that invoke unobservable entities. His procedure was simply to introduce a third partition $\{T\}$ that mediates between the previously mentioned partitions $\{S\}$ and $\{M\}$. In looking at the information transmitted by the system comprising $\{S\}$ and $\{T\}$, and that transmitted by the system comprising $\{T\}$ and $\{M\}$, it appears that the introduction of the 'theoretical' partition results in an increase in information. The scare-quotes on the word "theoretical" in the preceding sentence are designed to call attention to the fact that there is nothing intrinsically theoretical or unobservable about the partition in terms of T in Greeno's treatment. The same information-theoretic relationships would hold if all of the Ts referred to directly observable entities. Greeno's aim, however, was to provide a formal framework in which unobservables could fulfill an explanatory function.

Greeno's simple extension of his basic information-theoretic approach into the domain of theoretical explanation would have been beautiful, if it had worked, but as I tried to show in my contribution to the same symposium (Salmon, 1971), the apparent gain in information turns out to be illusory. In addition, the third symposiast, Jeffrey (1971), raised a number of difficulties regarding the suitability of transmitted information as a measure of explanatory force. Hanna (1978) also criticizes Greeno's choice of that measure.

For reasons that have already been given, at least in part, I am convinced that no philosophical theory of scientific explanation can be adequate unless it incorporates, in a central position, causal and theoretical factors. It seems clear that Greeno's (1971) attempt to accomplish this aim was not successful. The remaining chapters of this book will be devoted to a discussion of causal and theoretical aspects of scientific explanation. When these considerations are available to us, we will be in a better position to ascertain whether information theory can furnish a viable account of scientific explanation. Let me remark, by way of anticipation, that I think it stands a reasonable chance of doing so if it can furnish a serviceable concept of probabilistic causality. In view of such arguments as were brought forth by Kenneth Sayre (1977), there seems to be a strong possibility that the information-theoretic approach is equal to that task.

The Erotetic Version

Early in the first chapter, I remarked that although not all why-questions are requests for scientific explanations and not all requests for scientific explanation are made by posing why-questions, it seems possible to re-phrase any request for a scientific explanation as a why-question. This fact—if it is a fact—suggests a strategy for developing a philosophical theory of scientific explanation. Any request for a scientific explanation should be recast in some canonical form as a why-question, and scientific explanations should be analyzed as answers to why-questions. The logic of questions is known as *erotetic* logic; consequently, I am calling this version of the epistemic conception the "erotetic version." The approach is epistemic because the why-question is a request for something to fill a gap in someone's knowledge; the adequacy of an explanation is judged in terms of the manner in which the intellectual lacuna is filled. The epistemic character of this approach is further revealed by the fact that, viewed in this way, explanations must be relativized to knowledge situations. Whether a given explanation is satisfactory will invariably depend upon the knowledge-gaps a given individual needs or wants to bridge.

Development of the erotetic approach has stemmed to a large extent from Sylvain Bromberger's pioneering "Why-Questions" (1966). He carefully spells out such notions as the standard form of an explanation-seeking why-question, the presupposition of a why-question, and the conditions under which why-questions arise. In working out a suitable theory of why-questions, Bromberger sought to deal with various difficulties he saw in Hempel's theory of deductive-nomological explanation. The scope of his investigation was thereby restricted to explanations that appeal to universal laws; he did not undertake to provide any treatment of statistical explanation—an endeavor that is of great concern to us. Moreover, a critique by Paul Teller (1974) showed convincingly that Bromberger's theory did not accomplish its intended goal even with respect to the domain of D-N explanation.

An erotetic theory of scientific explanation that is at once broader and more sophisticated than that of Bromberger has been offered by van Fraassen (1980, chap. 5).[12] It deals in detail with statistical explanation, and it benefits from the work of Belnap and Steel, *The Logic of Questions and Answers* (1976), which was not available to Bromberger. We must take a serious look at van Fraassen's theory. I shall first attempt to provide a characterization of his account (van Fraassen, 1980, chap. 5, sec. 4), and then turn to a critical discussion of it.

What, according to van Fraassen, is a scientific explanation? It is an *answer* to a why-question. If this characterization is to be illuminating and applicable, we need to know with reasonable precision what why-questions and their answers are. It is important to be clear at the outset that questions and answers are abstract entities—they are *not* sentences in any language. The linguistic entity that corresponds to a question is an interrogative sentence; an interrogative sentence poses a question, but the question is not identical with the sentence. In this respect, questions are abstract entities in the same way that propositions are abstract entities. Different sentences may express the same proposition; two different tokens of the same sentence-type may, depending upon context, express different propositions. As van Fraassen points out, the sentence-type "I am here" always expresses a true proposition, but it expresses different true propositions depending upon the reference, in the context, of the token-reflexive terms "I" and "here." Similarly, when an interrogative sentence is uttered, we need to determine, by taking contextual factors into account, what question is being expressed. This is a general feature of questions; it obviously applies to the more limited class of why-questions.

Let us suppose that we have a request for an explanation; one of Hempel's standard examples will do. Whatever the actual words used to express the request, we translate it into canonical form. "Like, hey, man, how come that fire matches my canary?" becomes "Why did the Bunsen flame turn yellow?" The standard form is simply

Why (is it the case that) P_k?

where P_k is some proposition. P_k is called the *topic* of the question. In addition, according to van Fraassen, the question involves a *contrast class* X where

$$X = \{P_1, \ldots, P_k, \ldots\}$$

and the topic P_k is an element of X. In the Bunsen flame example, the contrast class might contain—in addition to the topic, the flame turned

---

[12] Raimo Tuomela (1981) offers another account of scientific explanation that falls within the erotetic version of the epistemic conception. In this discussion, I shall focus attention upon van Fraassen's account, for it seems better developed and more satisfactory than Tuomela's.

yellow—the alternatives: it turned green, it turned red, it turned purple, it remained blue, and so forth. It is important to emphasize that one and the same interrogative sentence may involve different contrast classes. The contrast class might instead include: the candle flame turned yellow, the flame in the kerosene lamp turned yellow, the flame of the blowtorch turned yellow, the flame on the gas stove turned yellow, and so forth. One has to look for contextual cues—for example, the emphasis placed upon the words—to determine what contrast class is intended. "Why did the Bunsen flame turn *yellow*?" would indicate the former contrast class, while "Why did the *Bunsen flame* turn yellow?" might suggest the second. The physical surroundings could also provide clues. If there are a bunch of different flames burning in the vicinity, that might reinforce the second choice.

In addition to the topic P_k, and the contrast class X, the why-question includes a *relevance relation R*. A proposition A (an answer) is considered *relevant* to the question if A bears the relation R to $<P_k,X>$. There are, it seems to me, many sorts of candidates for the relevance relation. R might be a causal relation; for example, the car went out of control because it aquaplaned on the wet highway. R might have a functional character; for example, the elephant has large ears in order to dissipate body heat. R might be a deductive relation; given the knowledge (part of the context) that pennies are made of copper and that copper conducts electricity, the fact that a penny was inserted behind the fuse entails that current will flow in a circuit that contains a blown fuse. The context may determine, for example, whether a causal or a functional explanation is desired. One's basic philosophical views about the nature of scientific explanation will have a large bearing upon one's views about admissible types of relevance relations R.

On van Fraassen's account, the foregoing three items determine a why-question, so we can conveniently identify the why-question Q with the ordered triple consisting of them (p. 143):

$$Q = <P_k,X,R>$$

Given this explication of why-questions, it is now possible to characterize their answers. We may begin by inspecting the canonical sentence that expresses a *direct answer D* to the why-question Q:

D: P_k *in contrast to* (the rest of) X *because A.*

The sentence D must express a proposition, and that proposition is determined by the same context that determined which why-question Q is expressed by the interrogative sentence "Why P_k?" The proposition expressed by D makes four assertions (p. 143):

1. P_k is true.
   (The fact-to-be-explained actually obtains; the event-to-be-explained actually occurred.)
2. For all P_j in X, if $j \neq k$, P_j is not true.
   (When we ask, according to van Fraassen, why F *rather than* G, we imply that G does not obtain. This easy gloss contains an immense pitfall, as I shall try to show later.)
3. A is true.
   (Since D contains *Because A*, which entails A, to assert D involves the assertion of the truth of A. It should be noted, however, that van Fraassen rejects the requirement of truth for explanations (p. 98), and places considerable emphasis upon the distinction between accepting A—for purposes of explanation or for other purposes—and believing A to be true (e.g., pp. 151–152). Hence, to accept D as an explanation of P_k does not require one to believe that A is true.)
4. A bears the relevance relation R to $<P_k,X>$.
   (When we say *because A*, we are claiming that A is explanatorily relevant to the explanandum.)

On the basis of the foregoing considerations, van Fraassen offers the following formal definition (p. 144):

B is a direct answer *to question Q* $= <P_k,X,R>$ *exactly if there is some proposition A such that A bears relation R to $<P_k,X>$ and B is the proposition which is true exactly if (P_k; and for all i $\neq$ k, not P_i; and A) is true.*

The proposition A is called the *core* of answer B, since the answer can be given by the abbreviated sentence "Because A." If we say, as I take it van Fraassen intends, that the entire direct answer constitutes a scientific explanation, then A by itself may be identified as the explanans.

Questions have presuppositions, and why-questions are no exception. If we are unwilling to accept some of the presuppositions, we consider the question inappropriate. One does not give an answer to an inappropriate question; one rejects the question. According to van Fraassen's account, a why-question has the following presuppositions (pp. 144–145):

1. Its topic is true;
2. In its contrast class, only its topic is true;
3. At least one of the propositions that bears its relevance relation to its topic and contrast-class is also true.

Items 1 and 2 constitute the *central presupposition* of question Q. We may say that a why-question Q *arises in context K* (assuming K to be logically consistent) if K implies the central presupposition of Q—items 1 and 2—and K does not imply the denial of any presupposition of Q—that is, K does not imply that the why-question has no answer. The issue of the circumstances under which a why-question arises has central importance for van Fraassen, inasmuch as he regards the problem of the rejection of requests for explanations as a serious obstacle for most accounts of scientific explanation. We reject a why-question by saying that it does not arise in the context. In Aristotelian physics, it is often said, we can ask for an explanation of the uniform motion of a body; whereas in Newtonian physics, we can ask for an explanation only of a change of motion (acceleration). In this latter context, the question of explaining uniform motion does not arise.

The foregoing exposition provides, I believe, an account of van Fraassen's formal machinery that will be adequate for our discussion. We have seen how he understands such concepts as scientific explanation, why-questions, presuppositions of why-questions, and direct answers to why-questions. We must now turn to the matter of evaluating proffered explanations—that is, judging answers to why-questions (sec. 4.4).

Three main considerations should be brought to bear in evaluating an answer. First, the proposition A—the core of the answer—must itself be weighed. If the knowledge context K entails that A is false, then we reject "Because A" as an answer. If the answer is not ruled out on that ground, then we must consider the probability that K bestows upon A.

Second, we must take account of the degree to which A *favors* the topic of Q in comparison with other members of Q's contrast class. This concept of favoring is not precisely defined by van Fraassen, but I shall not complain about that, since he gives a clear enough idea of what he is driving at.

Third, the answer "Because A" must be compared with other possible answers to our why-question in three distinct ways: (a) Is A more probable, relative to K, than other possible answers? (b) Does A favor the topic of Q more strongly than do other possible answers? (c) Do any other answers make A irrelevant to (i.e., screen it off from) the topic of Q?

The strongest case of an answer A favoring the topic P_k would seem to be that in which A taken in conjunction with the background knowledge K entails P_k and also entails the falsity of every P_j for which $j \neq k$. As van Fraassen realizes, this sort of characterization gives rise to an old puzzle about what is to be taken as comprised within that knowledge situation. The most usual situation in which we ask for an explanation of P_k is one in which we already know that the explanandum-statement is true and that the other members of the contrast class are false. Typically, then, the foregoing items follow trivially from K. What we want to do is to pare down the background knowledge by excluding those items of information, but without eliminating too much more. It is not easy to see how to characterize the circumscribed body of knowledge $K(Q)$ to which we may legitimately appeal in providing the direct answer to Q, which is the explanation of P_k. "Neither the other authors nor I can say much about it [how to excise items from K]," van Fraassen remarks. "Therefore the selection of the part $K(Q)$ of K that is to be used in the further evaluation of A, must be a further contextual factor" (1980, p. 147). In the context of statistical explanation, the problem at hand is none other than the problem of choosing suitable reference classes. Hempel dealt with this question in his (1962, 1965a, and 1968), and he endeavored to resolve it by means of his requirement of maximal specificity. I discussed the homogeneity of reference classes in (Salmon et al., 1971) and again in (Salmon, 1977). These efforts on the part of Hempel and me were not successful. I hope that the treatment of objectively homogeneous reference classes in the preceding chapter is more satisfactory. It is clear that van Fraassen is incorrect in claiming that we cannot say much about this subject; it is not clear that he would be wrong if he were to claim that we cannot say much that is worthwhile about it.

Let us suppose, in any case, that we have somehow isolated the portion $K(Q)$ of K that can be used in evaluating the answer A. We can then say that "A receives in this context the highest marks for favoring the topic $[P_k]$" (p. 147) if A and $K(Q)$ together imply the truth of the topic and the falsity of all other items in the contrast class X. This case bears some resemblance to Hempel's D-N explanation, but there are important differences. Hempel's D-N explanations are neither epistemically relativized (in his sense) nor context dependent (in van Fraassen's sense). In a proper D-N explanation, the explanans entails the explanandum without appeal to background knowledge. Thus each D-N explanation of a particular event must include, in the explanans, at least one law-statement and at least one particular statement of initial conditions. For van Fraassen, in contrast, either the law or the initial conditions might be part of the background knowledge $K(Q)$, and thus, not part of the explanation. Consider once again the well-known why-question about the color of the Bunsen flame. In a context in which the asker knows that the flame always turns yellow if a substance containing sodium is introduced, the appropriate answer will be that a piece of sodium chloride was placed in the flame. The explanation does not contain a law, but the background knowledge does include a law that can—to use Scriven's term—serve in a role-justifying capacity. Van Fraassen thus departs, in a way that bears some resemblance to Scriven's approach, from the covering-law conception of scientific explanation. In another context, the questioner may be aware that a piece of rock salt has been placed in the Bunsen flame, but is ignorant either of the fact that rock salt is a sodium compound or that sodium turns Bunsen flames yellow. In that case, an adequate explanation will contain one or more generalizations, but will not include the statement of initial conditions.

In situations in which A does not, in conjunction with $K(Q)$, entail the truth of Q's topic and the falsity of all other members of its contrast class, we must judge the manner in which A favors the topic in terms of the way in which it redistributes the probabilities among the members of the contrast class. This matter must be approached from the standpoint of a distribution of prior probabilities, on the basis of $K(Q)$, of the items P_j in the contrast class X. Adding A to $K(Q)$, we get a distribution of posterior probabilities of the P_j's conditional upon A. As I understand van Fraassen's basic idea, A favors P_k to the extent that A makes P_k stand out to a greater degree as expectable vis-à-vis its fellow members of X. As van Fraassen explicitly notes (p. 148), A may favor P_k even if A lowers its probability. Consider a concrete example. Suppose there will be a horse race in which there are eight entries. On the basis of prior information contained in the racing form, we may assign the following prior probabilities of winning:

Horse #1, 0.35
Horses #2–#3, 0.3 each
Horses #4–#8, 0.01 each

On the day of the race we find out that there are special track conditions that improve tenfold the chances of horses #4–#8, while lessening the chances of horses #1–#3, with the result that the posterior probabilities are:

Horse #1, 0.3
Horses #2–#8, 0.1 each

Although the information about the track conditions lowers from 0.35 to 0.3 the probability that horse #1 will win, it favors the proposition that horse #1 wins, for the conditions widen the gap between horse #1 and its closest contenders.

With this informal understanding of the concept of favoring, let us examine the criteria that van Fraassen proposes for the evelution of answers to why-questions. The first question, it will be recalled, concerns the acceptability of the explanans A. In considering this point, it is important to distinguish two different cases. We are sometimes concerned, as Hempel would phrase it, with *potential explanations*—that is, we want to know how satisfactory an explanation A would provide if it were true, but we are not at the moment concerned to ascertain the truth-value of A. In this case, van Fraassen's first criterion is beside the point. In the second case, we want to know whether A provides an *actual explanation*. With Hempel,

I would be inclined to say that the truth of *A* is a necessary condition for it to furnish an actual explanation of the explanandum (the topic of *Q*). Of course, if we are not prepared to accept *A* as true (or, perhaps, likely to be true), we will not be willing to accept it as an actual explanation of anything.

The second criterion involves favoring. Clearly, in comparing the prior probability distribution with the posterior probability distribution, van Fraassen is looking at statistical relevance relations; in admitting that an explanation can sometimes involve a relation of negative relevance between the explanans and the explanandum, he exhibits some strong affinities with the statistical-relevance approach. His later reference to the screening-off relation reinforces this impression.

It is in this criterion, however, that van Fraassen and I find our main parting of the ways. In a parenthetical remark on his second criterion, he says, "This is where Hempel's criterion of giving reasons to expect, and Salmon's criterion of statistical relevance may find application" (1980, p. 146). At this juncture, I think, we are forced to make a fundamental choice between the epistemic conception and the ontic (or causal) conception of scientific explanation. Van Fraassen seems to choose the epistemic route, for his criterion of favoring *does* involve reasons to expect. Although there is not, strictly speaking, a simple ordering, one can discern roughly a weakening series of sorts of reasons to expect. Those who reject all non-deductive kinds of explanation (e.g., such proponents of the modal conception as von Wright) regard deductive certainty as the only kind of reason to expect which is strong enough for scientific explanation. When Hempel adopted his inductive-statistical model, he allowed high inductive probability as a sufficiently strong type of reason to expect. In my first critique of Hempel's I-S model (Salmon, 1965), I proposed positive statistical relevance; this is a still weaker sort of reason to expect. L. J. Cohen (1975) maintains that I should not have relinquished it as a requirement for acceptable explanations, and Tuomela (1981, p. 277) explicitly rejects the notion that relations of negative relevance can have any explanatory import. It was most ingenious of van Fraassen to have devised the notion of favoring, which admits cases of negative relevance, but still provides grounds for an increased expectation in a comparative sense. It is hard to see how one could weaken the reason-to-expect criterion any farther without doing it in completely.

It is precisely at this point—van Fraassen's second criterion for the evaluation of explanations—that a fundamental disagreement arises between us. Let us accept, at least for purposes of discussion, the erotetic machinery of why-questions, presuppositions, contrast classes, and direct answers. For reasons which have been spelled out previously in connection with the inferential version of the epistemic conception, I am strongly inclined to reject the notion that only favored members of the contrast class can be explained. Let us return to some familiar examples. In the Mendelian experiment on the color of pea blossoms, we have a contrast class X containing: the blossom is red, the blossom is white, the blossom is some other color. Pea blossoms come in a variety of colors, but in the population of the particular experiment the probabilities are about 3/4 for red, about 1/4 for white, and (allowing for sports) a very small nonzero value for some other color. Regardless of the color of the particular blossom we single out for explanation—regardless of whether the topic is *this blossom is red* or *this blossom is white*—the contrast class is the same. The background information—containing Mendel's genetic theory—is also the same in both cases, and so is the explanans A, which specifies the genetic character of the population. Consequently, both the prior and posterior probability distributions are the same. Thus, in either case, the topic *this blossom is red* is favored. On van Fraassen's theory, then, the explanation of the color of the red blossom (which favors the topic of its question) is clearly superior to the explanation of the color of the white blossom (which does not favor the topic of its question). Nevertheless, as I remarked previously, I am firmly convinced that, in such cases, we understand the unfavored outcomes just as well (or as poorly) as we understand the favored ones. Similar remarks obviously apply to other examples—such as suicide, delinquency, and radioactive decay—which have been introduced in an effort to illustrate the point that we understand the improbable outcome just as well as the probable outcome in any type of situation that probabilistically gives rise to different outcomes with differing probabilities. This is one of the fundamental points that was elegantly argued by Jeffrey in (1969).

Van Fraassen's third criterion for the evaluation of answers involves three distinct comparisons of a given answer A with other possible answers to the same question. First, we must ask whether other possible answers are more probable relative to K. If A were significantly less probable than some other answer, that would be a serious mark against it. Second, we must ask whether other answers favor the topic more than A does. I find this part of the third criterion unacceptable for exactly the same reason brought against van Fraassen's second criterion, namely, because it appeals to the notion of favoring. Third, we must ask whether A is rendered irrelevant by some other answer. This use of the screening-off relation seems admirable.

Although van Fraassen's contrast class—which is the same as Greeno's explanandum-partition—constitutes an important component in scientific explanation, we must now call attention to a serious danger associated with its use. In the first instance, the contrast class is used to clarify the explanation-seeking why-question. For this purpose it appears to be a useful tool. Next, it shows up in the presupposition asserting that the topic of the why-question is the only member of the contrast class that is true. The presupposition is, I believe, sound. This presupposition seems, however, to lead almost imperceptibly to what I take to be a fundamental error— namely, to the supposition that in giving an explanation, we have explained why P_k *rather than* P_j $(j \neq k)$ obtains. To see how easily this transition occurs, consider van Fraassen's gloss on that very presupposition: "It does not make sense to ask why Peter *rather than* Paul has paresis if they both have it" (p. 143, italics added). But we should note carefully that van Fraassen's why-question Q does not ask "Why P_k rather than P_j?" The why-question asks "Why P_k?" and in so doing it implies that P_j $(j \neq k)$ is false. When we receive the answer, "Because A," we have been told *why* P_k; we have *not* necessarily been told *why*, but only *that*, P_j is false. This point may be subtle, but it is important.

Grant me, at least for the moment, that we have some genuinely in- deterministic chance device—for example, an electron that approaches a potential barrier of such a kind that it has a 0.9 chance of being reflected back and a 0.1 chance of tunneling through. Suppose it is reflected. We formulate the question "Why was the electron reflected by the barrier?" and we identify the contrast class as {it is transmitted, it is reflected}. Now, according to a great many theories of explanation, we can give the desired explanation by describing the initial conditions of the experiment and, perhaps, citing the pertinent statistical laws. Thus, we can say, the electron was reflected because of the way it impinged upon the barrier and because of a lawfully determined probability, under those circumstances, of that outcome. The statement implies, of course, that the electron was—as a matter of fact—reflected and not transmitted in the particular case at hand. The answer does not explain why it was reflected rather than transmitted in this case; it just happened that way, as it does in 9/10 of such cases. In 1/10 of such cases, it just happens that the electron tunnels through and is not reflected. But, by hypothesis, there is no explanation for the fact that it does one of these on one occasion and not on another. If we assume that an explanation of why one outcome occurs must *ipso facto* be an explanation of why one rather than another occurs, we run a serious chance of finding ourselves involved in the notion that only those events that are strictly determined can be explained. The slip into the *rather than* mode runs a great risk of carrying us out of the epistemic conception and into the modal conception. This moral of the story will, I think, become much clearer as we address the modal conception—the subject to which we shall now turn. I shall have more to say about van Fraassen's approach to scientific explanation when the three general conceptions are compared at the close of this chapter.

THE MODAL CONCEPTION

According to the modal conception, the aim of scientific explanation is to show that an event, which at first blush looks as if it might or might not have occurred, in fact had to occur. The explanation renders the explanandum-event physically necessary relative to the explanatory facts. What happens to this conception when we make the transition to the indeterministic context? The most straightforward answer, in my opinion, was given by von Wright, who simply denies the possibility of statistical explanation:

> It is part and parcel of an inductive-probabilistic explanation that it admits the possibility that E might have *failed* to occur. It therefore leaves room for an additional quest for explanation: why did E, on this occasion, actually occur and why did it not fail to occur? It would be the task of deductive-nomological explanation to answer *this* question. Sometimes we can answer it . . . [but] failing such additional information which gives us a deductive-nomological explanation of E, we have not explained why E occurred, but only why E was to be expected. . . . It seems to me better, however, not to say that the inductive probabilistic model explains what happens, but to say only that it justifies certain expectations and predictions. (1971, p. 13)

Von Wright has, it seems to me, correctly perceived the consequences that must be drawn from the modal conception with respect to statistical explanation. Since I find the consequences unacceptable, I think we must reject the modal conception. Quite possibly the greatest success story of modern science, with respect to its capacity for explaining a wide range of phenomena, is that of quantum mechanics. Under the most widely accepted interpretation, quantum mechanics is a statistical theory, and the explanations it provides are statistical explanations. The modal conception cannot survive the transition from the deterministic context to the indeterministic context.

D. H. Mellor, who—it will be recalled—gives a modal characterization of causal explanation, takes a different tack when he considers the indeterministic context. Just after stating that "[c]ausal explanation closes the gap by deducing what happened from known earlier events and deterministic laws," he goes on to say:

Sometimes, however, suitable causal explanation is not to be had. . . . we cannot . . . close the gap that calls for causal explanation. But perhaps we can narrow it. This epistemic possibility of an explanandum's falsehood comes by degrees, and relative probability I take to be (*inter alia*) the measure of it. So gaps that causal explanation would close completely may be partly closed by probabilistic explanation; that indeed I take to be its object. That being so, it is the better, *ceteris paribus*, the less epistemic possibility it leaves the explanandum's falsehood; *i.e.*, the more it raises the explanandum's probability relative to the complete explanans. (1976, p. 235)

From a strictly modal standpoint, it seems to me, we can distinguish necessity, impossibility, and possibility. Given a causal explanation, as Mellor construes it, we have the necessity of the explanandum and the impossibility of its failure to obtain. Where causal explanation, in his sense, cannot be achieved, we have the possibility of the explanandum's occurrence and the possibility of its nonoccurrence. At this point von Wright says, if I understand him correctly, that that is as far as the modal conception can take us. Mellor wants to go a step farther by introducing *degrees of possibility*; he does so by taking probability as a measure of degree of possibility. It is not quite clear to me how the identification of probability with degree of possibility enhances our understanding of the matter. One approach, which Mellor kindly conveyed in a private communication, exploits the logical interpretation of probability—an interpretation which, incidentally, Mellor rejects but Hempel adopts. It is widely agreed that deductive entailment represents a basic type of necessity. A number of probability theorists, including Carnap, have sometimes used the notion of partial entailment to give an intuitive feeling for the relation of degree of confirmation or degree of inductive probability. Employing this idea, one might then say that the degree to which premises inductively support—that is, partially entail—a conclusion is the *degree* to which those premises *necessitate* that conclusion. This invocation of the notion of partial entailment as a way of understanding logical probability strikes me as unenlightening, for, as I tried to argue in a rather extended study of partial entailment (Salmon, 1969), there seems to be no nonarbitrary way of assigning numerical values to degrees of partial entailment. One is, of course, free to make definitional stipulations about degrees of necessity and degrees of partial entailment, but such moves do not improve our understanding of explanation or confirmation. I think it is misleading to suppose that we have thereby made sense of such concepts as *almost necessary* or *nearly impossible*. Once we admit an exception, we have to give up necessity altogether.

It seems to me preferable—though Mellor would take strong exception to this approach—to introduce probabilistic causation. With such concepts we can talk intelligibly about the strength of the tendency of certain circumstances to yield an occurrence of a given sort; we can measure the degree of strength of a tendency without getting involved in degrees of possibility.[13] This is the view I shall attempt to develop in chapter 7. If the matter is handled in this way, Mellor's approach to statistical explanation falls properly within the ontic (or causal) conception. There remains a basic difference, however, between Mellor's version and mine: he wishes to retain the high-probability requirement (1976, pp. 232, 241), while I do not. The retention of that requirement, along with the introduction of degrees of possibility, seems to me to reflect a lingering desire for Laplacian determinism, or if worse comes to worst, as close an approximation thereto as possible.

The modal conception can be bolstered by appeal to a compelling intuition that we seem to have inherited from classical physics and the Laplacian determinism that it embodies. I believe that it is an anachronistic carry-over from that context, and that, difficult as it may be, we must shake it off if we are to come to terms with explanation in contemporary science—that is, with statistical explanation. Let me begin by stating the principle in its most minimal form, one which the majority of philosophers and scientists would probably accept as an indispensable condition of adequacy for any account of scientific explanation—indeed, I do not know of anyone who has heretofore had the temerity to come right out and reject it. This pivotal principle is:

1. *If a given set of facts provides an adequate explanation of some event*
   E, then those same facts cannot suffice to explain the nonoccurrence
   of E.

This is Mellor's adequacy test S (1976, p. 237). Hempel's deductive-nomological pattern of explanation conforms to principle 1, for if the statement that *E* occurs follows deductively from a consistent set of premises, the statement that *E* fails to occur cannot also follow from the same premises. Prior to 1977, when he relinquished the high-probability re-

---

[13] Fetzer (1981) develops an account of statistical explanation in terms of a propensity theory of probability. In regarding probabilities as dispositions, and in rejecting the view that laws and dispositions can be adequately explicated in extensional terms, he appears to opt for a modal conception. In elaborating his view, he makes explicit use of the notion of partial entailment without giving any clear indication of how that concept should be understood. In (Salmon, 1969), I have spelled out in detail some of the profound difficulties that attend that concept. It does not appear that Fetzer has given any indication of how he hopes to circumvent these problems.

quirement for inductive-statistical explanation, the I-S model also conformed to it, for if the occurrence of E has a probability greater than 1/2 on given evidence, then the probability of non-E must have a value less than 1/2 relative to that same evidence. If the high-probability requirement is dropped, however, it is not easy to see how principle 1 can be retained.

It may seem that principle 1 is too evident to require defense. If, nevertheless, we request a reason for adopting it, we may be told that it serves to block certain sorts of pseudo-explanations—for example, 'explanations' in terms of a supernatural will or goal that is so vaguely specified that whatever happens may be said to conform to it.[14] We must, of course, exclude such 'explanations' from science, but I do not think we need principle 1 to accomplish that end. It is sufficient to require that statements of alleged explanatory 'facts' be reasonably well confirmed before they are asserted in explanatory contexts—or at the very least, that they be open to scientific confirmation or disconfirmation. This requirement is *not* tantamount to a verifiability criterion of scientific meaning; it is merely the requirement that scientific assertions be supported by evidence. I am not committing logical positivism.

Freudian psychoanalytic theory is often cited as another rich source of examples of vacuous 'explanations' that are to be rejected on the basis of principle 1. For instance, a boy's aggressive behavior toward his father may be 'explained' in terms of his unconscious hatred of his father, but apparently affectionate and solicitous behavior might be 'explained' by the same hypothesis. Fear of revealing the unconscious hostility, it is said, may lead to a reaction formation, and to behavior that is just the opposite of hostility. Another example is Freud's analysis of certain dreams, which appeared to fulfill no wish, as dreams that fulfill the wish to refute Freud's theory that all dreams are wish fulfillments. In (Salmon, 1959), I argued that in each of the foregoing cases the psychoanalytic hypothesis is not compatible with every conceivable phenomenon; but if it were, it would, of course, be lacking in explanatory value. Under these circumstances, however, we would not need to invoke principle 1; it would again suffice to reject such hypotheses on the basis of their immunity to all possible confirmation or disconfirmation.

In "Aspects of Scientific Explanation" (1965a), Hempel imposed a somewhat more stringent requirement upon I-S explanations, namely, "Any statistical explanation for the occurrence of an event must seem suspect if there is the possibility of a logically and empirically equally sound prob-

---

[14] An example that fits this description is the telefinalist hypothesis of Pierre Lecomte du Nouy in his *Human Destiny* (1947). Another example, somewhat more timely, occurs in so-called creation science.

abilistic account for its nonoccurrence'' (p. 395). This principle might be formulated as follows:

2. *It must not be possible that, when a given set of facts provides an adequate explanation of some event E, any other facts obtain that would constitute an adequate explanation of non-E.*

Hempel introduces his *requirement of maximal specificity* to secure adherence to principle 2. In the absence of some such requirement, he feels, we may find ourselves in an unhappy situation: "It is disquieting that we should be able to say: No matter whether we are informed that the event in question (e.g., warm and sunny weather on November 27 in Stanford) did occur or that it did not occur, we can produce an explanation of the reported outcome in either case; and an explanation, moreover, whose premises are scientifically established statements that confer a high logical probability upon the reported outcome" (p. 396). When, however, the requirement of maximal specificity is imposed, "We are *never* in a position to say: No matter whether this particular event did or did not occur, we can produce an acceptable explanation of either outcome" (p. 400).[15] When the high-probability requirement is given up, however, principle 2 need no longer be satisfied.

Principles 1 and 2 do not, without further assumptions, lead to the modal conception of scientific explanation, which may be formulated as follows:

3. *If, under a set of circumstances of type C, an event of type E sometimes occurs and sometimes fails to occur, then the fact that circumstances of type C obtained cannot constitute an adequate explanation on a given occasion of the occurrence of an event of type E.*

Proponents of the modal conception might insist upon a stronger formulation:

4. *If, under a set of circumstances C, it is possible for event E to fail to occur, then the fact that circumstances C obtained cannot constitute an adequate explanation of the occurrence of E.*

Harré and Madden express this view when they assert, "And if something else than what must happen could happen within any given system, then no explanation could ever be possible within that system, because one would not have succeeded in explaining the occurrence of one event rather than the other" (1975, p. 39). Principle 4 is stronger than principle 3, but

---

[15] See (Hempel, 1968) for further discussion and refinement of his requirement of maximal specificity.

since principle 3 would effectively block all statistical explanations of particular events, I shall direct my discussion toward it.

There are two grounds on which one might argue for principle 3. One of these has been offered by Stegmüller (1973, p. 284), who invites us to consider a situation in which E happens in 98% of all cases in which circumstances C obtain, and fails to occur in 2% of such cases. When E does occur, we may be tempted to explain it in terms of circumstances C; we must note, however, that in those cases in which E fails to occur, there cannot be any explanation of E—we cannot explain an 'event' that does not occur. But given circumstances C, it is a matter of chance whether E occurs or does not; therefore, it is a matter of chance whether C is an explanation of anything or not. This conflicts with our intuitions according to Stegmüller. If, however, we admit explanations of low-probability events, then we can say that sometimes C explains the occurrence of E and that sometimes C explains the nonoccurrence of E. This move, however, violates principle 1, which Stegmüller is unwilling to sanction.

The second argument for principle 3 is based upon symmetry considerations already introduced in connection with explanations of low-probability events. In discussing the example of genetic experiments on the color of pea blossoms, I argued that we understand the occurrence of a white blossom (probability 1/4) just as well as we understand the occurrence of a red blossom (probability 3/4). These symmetry considerations show, it might therefore be argued, that if we admit any statistical explanations at all, then we must admit low-probability explanations. But if we allow low-probability explanations, then precisely the same circumstances will explain the occurrence of red and also the occurrence of nonred blossoms. This violates principle 1. Since principle 1 is so dear to our intuitions, it may be argued, we must accept principle 3 and block *all* statistical explanations of particular events.

It should, perhaps, be mentioned at this point that adherents of the modal conception of scientific explanation will find no aid or comfort in Railton's (1978) so-called deductive-nomological model of probabilistic explanation (D-N-P model). One obviously cannot deduce from statements of particular initial conditions and a probabilistic law that a particular explanandum-event E occurs, and Railton does not claim otherwise (1978, p. 216). It is possible, at most, to deduce that E has a certain probability in the circumstances. It does not follow from this conclusion that E *occurs with a certain probability p*, for E may simply fail to occur in any given instance. The statement that E actually occurred must be attached as a "parenthetic addendum" (p. 214) that is true *by chance*.

One alternative remains open to those who wish to adhere to the modal conception—or to principle 1, at least—and to admit some form of statistical explanation. Hempel has distinguished two types of statistical explanation: inductive-statistical, in which a particular event is explained by inductive subsumption under laws, at least one of which is statistical; and deductive-statistical, in which a statistical regularity is explained by deductive subsumption under other laws, at least one of which is statistical. It is possible to maintain that there are no such things as I-S explanations of particular events, but only D-S explanations of statistical regularities. Thus, it might be argued, in the experiments on pea blossoms, one can explain the distribution of 1/4 white and 3/4 red, but not the color of an individual blossom. Similarly, in various quantum mechanical examples, such as radioactive decay, the statistical laws enable us to explain the frequency with which atoms of a certain type decay, but not the particular decays of individual atoms.

Although this approach has a certain appeal, I do not believe it is tenable. If, for example, we claim that Mendel's laws explain only a probability distribution, then we must forego all attempts to explain why a particular blossom has a particular color, but we must go much farther. We must also admit that we cannot explain why in a population of 500 plants, approximately 375 have red blossoms, for that outcome merely has a high probability, but is by no means a deductive consequence of the laws of genetics and the pertinent initial conditions. If one is willing to admit I-S explanations, provided they have high associated inductive probabilities, then the shift from the individual case to the large finite population will often handle the situation, but if no I-S explanations are to be allowed, then this approach is no more admissible for large groups of individuals than it is for the individual case all by itself.

It could, of course, be argued that theoretical science is under no obligation to provide explanations of individual occurrences or restricted groups of occurrences—that its entire concern is with regularities. Even though this view may, perhaps, have some initial plausibility, I do not believe it is correct. Consider another example from genetics. In an article entitled "Genes that Violate Mendel's Rules," James F. Crow reports on an experiment with fruit flies conducted in 1956 by his student Yuichiro Hiraizumi:

Hiraizumi mated hybrid red-eyed males . . . to females that . . . were white-eyed. Because there is normally no crossing-over . . . according to the rules of Mendelian inheritance half of the progeny should have had white eyes and the other half should have had dark red eyes. Hiraizumi did observe the expected 50:50 distribution among the progeny of some 200 matings. . . . Six of the matings, however, produced very strange results. Instead of half of the 100-odd progeny of each mating being red-eyed, almost all of them were. The red-eyed proportion ranged from 95-100 percent, and it was 99 percent or more in most cases.

What had caused six chromosomes, each descended from a different male . . . to behave in this most unusual way? (1979, p. 134)

Clearly, this is a theoretical question, and its answer (as Crow explains) involves an investigation of the chemical mechanisms of the 'cheating genes.'

The same sorts of remarks could be applied to important results in modern physics. For example, the famous electron diffraction experiments of Davisson and Germer gave rise to statistical distributions of scattered electrons that were quite puzzling and quite out of harmony with the results Davisson predicted on the basis of his theory. Early in 1925, after an interval in which Davisson had not been conducting his scattering experiments, he and Germer resumed them, but they shortly discovered that the apparatus had broken. Repairs had to be undertaken.

> By 6 April 1925 the repairs had been completed and the tube put back into operation. During the following weeks, as the tube was run through the usual series of tests, results very similar to those obtained four years earlier were obtained. Then suddenly, in the middle of May, unprecedented results began to appear. . . . These so puzzled Davisson and Germer that they halted the experiments a few days later, cut open a tube, and examined the target . . . to see if they could detect the cause of the new observations. (Gehrenbeck, 1978, p. 37)

We all know the outcome of the story. The pattern—that is, the statistical distribution—that Davisson and Germer obtained in 1925 bore some resemblance to diffraction patterns obtained with light and X rays. The explanation of these results was given in terms of de Broglie's wave theory of matter, as Davisson learned at a conference in Oxford in the summer of 1926:

> Davisson, who generally kept abreast of recent developments in his field but appears to have been largely unaware of these recent developments in quantum mechanics, attended this meeting. Imagine his surprise, then, when he heard a lecture by Born in which . . . his curves of 1923 were cited as confirmatory evidence for de Broglie's electron waves! (ibid.)

It might be argued, I suppose, that in each of these cases we are dealing with a large enough number of events to say that the observed distribution over the finite sample is taken to reflect the distribution that would be obtained in the long run. Thus, it might be said, it is the long-run distributions that are explained by the theory; the observed cases are *evidence for* the fact-to-be-explained, but they do not *constitute* the explanandum. I do not think this sort of response will handle all examples of statistical explanation. In Rutherford's famous scattering experiments with alpha particles, for instance, even a very small number of particles scattered at large angles would demand a theoretical explanation of the sort given by his nuclear model of the atom. Moreover, it seems to me, the melting of just one ice cube in a glass of tepid water demands a statistical explanation in terms of the molecular-kinetic theory of heat.

Even if the original thesis were true—though examples of the foregoing sorts seem to me thoroughly to undermine it—that theoretical science has no place for statistical explanations of individual events or finite collections of them, it would be essential to remember that the sciences have their applied, as well as their pure, aspects. In applied science, there can be little doubt that we are often concerned with the explanation of individual occurrences. We want to know why a particular dam collapsed, or why a particular airplane crashed, or why a particular soldier contracted leukemia, or why a particular teen-ager became delinquent. If some of these explanations employ statistical laws, that does not constitute an adequate reason for denying that they are explanations after all. Moreover, if philosophers of science insist upon dealing only with pure science, refusing to attend to its applications, they will be guilty of ignoring a major aspect of the scientific enterprise. We can conclude, I believe, that there are bona fide statistical explanations in the sciences that do not qualify as deductive-statistical. Such explanations must conform to the I-S or S-R models, or to some model closely akin to them.

I have not discussed the D-S model of explanation in any detail before now because it seems otiose to regard it as separate and distinct from the D-N model. D-S explanations are arguments that qualify as valid deductions, and they include at least one law-statement among their premises. The fact that D-N explanations, as characterized by Hempel, contain only universal laws, seems to be an inessential restriction; I would suggest that this restriction be dropped, and that the D-S model be incorporated within a slightly relaxed D-N model. Statistical laws have just as much right to the title "nomological" as do universal laws. In my view, of course, neither the explanations that Hempel would classify as D-N nor those that he would classify as D-S are to be characterized as arguments. They are, nevertheless, legitimate scientific explanations, and I would characterize them in terms of the *physical* subsumption relation discussed previously. In either case, whether the regularity-to-be-explained is universal or statistical, its explanation consists in showing that it fits into a pattern—one which is constituted by universal or statistical regularities in the world.

If science does contain bona fide explanations of the statistical sort (other than those Hempel classifies as D-S), then the modal conception of scientific explanation is untenable. As I tried to show in the discussion of van Fraassen's concept of favoring (pp. 105–107 of this chapter), we cannot go along with the demand—explicitly voiced by Harré and Madden—that every explanation show why one event *rather than another* occurred. This requirement, which is appropriate within the context of Laplacian explanation, is no longer appropriate to twentieth-century science. In the deterministic context of classical physics, each event that occurs is necessitated, and it is explained by showing that it fits into its necessary place in the physical pattern of events. If twentieth-century physics is correct, then—as most interpreters see it—we must accept irreducibly statistical laws, and we must look upon events as having places in stochastic patterns. If we can manage to relinquish the rather compelling assumption that explanations are arguments—inductive or deductive—then we need not demand deductive certainty (as the modal conception requires), high inductive probability (as the inferential version of the epistemic conception is willing to settle for), or even such weak sorts of expectation as positive relevance (L. J. Cohen, 1975) or van Fraassen's favoring (1980, pp. 147–148). It is enough to be able to assign the actual probability. If determinism is false, a given set of circumstances of type C will yield an event of type E in a certain percentage of cases; and under circumstances of precisely the same type, no event of the type E will occur in a certain percentage of cases. If C defines a homogeneous reference class—not merely epistemically homogeneous but objectively homogeneous—then circumstances C explain the occurrence of E in those cases in which it occurs, and exactly the same circumstances C explain the nonoccurrence of E in those cases in which it fails to occur. The pattern is a statistical pattern, and precisely the same circumstances that produce E in some cases produce non-E in others.[16] There are no relevant circumstances, in addition to C, that can be brought to bear to explain why E occurs on one occasion, and why E fails to occur on others. If we look at scientific explanation in this light, principle 1 can comfortably be abandoned, in Russell's phrase, as "a relic of a bygone age." We need have no fear that science will thereby be contaminated by vacuous 'explanations,' for the statistical law-statements to which we appeal are empirically testable by well-known statistical methods.

[16] It may be noted that the probability of E, given C, will often be different from the probability of non-E, given C, and that, therefore, the explanation of E is not the same in *every* respect as the explanation of non-E. I am not inclined to invoke that argument, however, since I want to leave open the case in which $P(E|C) = P(\bar{E}|C) = 1/2$.

THE ONTIC CONCEPTION

The aim of a scientific explanation, according to the ontic conception, is to fit the event-to-be-explained into a discernible pattern. This pattern is constituted by regularities in nature—regularities to which we often refer as laws of nature. Such laws may be either universal or statistical, and if they are statistical, they need not confer upon the event-to-be-explained a high probability. Among the events that fall under the statistical regularities that obtain in the world, some have high, some have middling, and some have low probabilities. That does not matter as far as this conception of explanation is concerned.

It should be immediately evident, however, that mere subsumption under laws—mere fitting of events into regular patterns—has little, if any, explanatory force. Early in the first chapter, it will be recalled, I cited the pre-Newtonian knowledge of the relationship of the tides to the position and phase of the moon as a prime historical example of subsumption of natural phenomena under regularities that was totally lacking in explanatory value. It was only when the Newtonian explanation of that regularity in terms of the law of gravitation became available that anyone could maintain plausibly that the tides had been explained. The obvious moral to be drawn from this example, and many others as well, is that some regularities have explanatory power, while others constitute precisely the kinds of natural phenomena that demand explanation. The two classes of regularities, of course, are not mutually exclusive; some regularities that possess explanatory power are themselves amenable to explanation at a 'higher' or 'deeper' level. If we want to elaborate an ontic conception that differs in any significant way from the various versions of the epistemic conception, we shall have to come to an understanding of the distinction between the two types of regularities. One idea that immediately occurs should be scotched without delay. The distinction we are seeking does not coincide with the distinction between lawful and nonlawful regularities. As we noted in chapter 1, the ideal gas law is, indeed, a lawful regularity, but until it is explained by something like the molecular-kinetic theory of gases, it does not have much (if any) explanatory import on its own. We are looking for a different distinction.

Most authors who have adopted the ontic conception have construed it in causal terms. The relationships that exist in the world and provide the basis for scientific explanations are causal relations. This general conception receives its most powerful initial support from paradigms—usually explanations of particular occurrences in contrast to general regularities—that are found in applied science or in everyday common-sense contexts. The basic idea is quite straightforward. To provide an explanation of a particular event is to identify the cause and, in many cases at least, to exhibit the causal relation between this cause and the event-to-be-explained. Thus to explain a particular airplane crash, an FAA investigator might find that it was due to fuel of the wrong type (e.g., jet fuel rather than gasoline) having been introduced into its fuel tanks. The explanation might be further elaborated by going into some of the details of the engine's operation, showing why it malfunctions when fuel of an inappropriate type reaches its cylinders.

The major shortcoming of this approach, as it is often presented by such proponents as Michael Scriven (1975) and Larry Wright (1976), is that no adequate analysis of causal relations is provided. This is, I believe, a serious drawback. Hume's incisive critique of causal concepts, which has led some philosophers to eschew causality altogether in their accounts of scientific explanation, cannot be merely brushed aside. Hume's analysis poses fundamental philosophical problems, and we must come to terms with them if we are to have any hope of providing an adequate explication of causal explanation. Moreover, even neglecting Humean problems, causal relations are subtle and complicated. We need to furnish analyses of causal processes and causal interactions, and to clarify their relations to one another. We must give an account of the causal explanation of regularities, both universal and statistical, in addition to our theory of causal explanation of individual events. We must take a careful look at the *principle of the common cause*. Reichenbach attempted to formulate this principle in his posthumous work, *The Direction of Time* (1956), and he gave some brief indications of its importance to the theory of scientific explanation. It has, however, been largely overlooked by most philosophers writing on scientific explanation, though van Fraassen is a notable exception. We shall have to remedy that omission. In examining Reichenbach's formulations, we shall find that they have certain defects that need to be repaired, but the revised version will play a key role in our treatment of explanation. It will be the business of the next three chapters to deal with all of these causal concepts, and to fit them into a theory of causal explanation. It is only by providing adequate treatments of such causal concepts, I believe, that we can successfully implement the ontic—or as some may wish to say—the causal conception of scientific explanation.

Although the causal conception seems altogether appropriate for a Laplacian kind of world, it might be thought that the idea of causal explanation would lead into difficulties when we go from the deterministic to the indeterministic context. No such consequence need follow, however, if we can provide a suitable probabilistic conception of causality. Various authors—principally I. J. Good (1961), Hans Reichenbach (1956), and Patrick Suppes (1970)—have attempted to elaborate probabilistic or statistical conceptions; more recently, Fetzer and Nute (1979) and Sayre (1977) have also taken up the task. While I do not consider the results completely satisfactory, I do not believe there is anything incoherent, absurd, or self-contradictory in the attempt to explicate such a notion. It makes good sense, in my opinion, to suppose that causal influences are propagated in a probabilistic fashion, and that causal interactions give rise to statistical distributions of results. Among the causal processes in the world, some, at least, are stochastic. One of the main tasks to be undertaken in the next three chapters—chiefly chapter 7—will be the elaboration of a more adequate probabilistic concept of causality.

Various authors who seem committed to some version of the ontic conception have adopted what might be called a *mechanistic* standpoint. In discussing such examples as coin-tossing and genetically determined traits, Jeffrey (1969) places considerable emphasis upon the role of the stochastic processes that lead to the various outcomes. These processes are the physical mechanisms that are responsible—probabilistically—for the phenomena we are trying to explain. Inasmuch as Jeffrey seems to reserve the concept of causality for deterministic situations, he introduces a different vocabulary for the stochastic situations. Jeffrey is entirely correct, I believe, in forcefully calling attention to the explanatory role of the physical mechanisms; nevertheless, with the aid of a suitable theory of probabilistic causality, it will be possible to handle all such cases within the causal framework.

In a discussion of such basic microphysical phenomena as spontaneous radioactive decay, and its explanation in terms of quantum mechanical tunneling, Railton (1978) adopts an even more explicitly mechanistic approach: "The goal of understanding the world is a theoretical goal, and if the world is a machine—a vast arrangement of nomic connections—then our theory ought to give us some insight into the structure and workings of the mechanism, above and beyond the capability of predicting and controlling its outcomes. . . . What is being urged is that D-N explanations making use of true, general, causal laws may legitimately be regarded as unsatisfactory unless we can back them up with an account of the mechanism(s) at work'' (1978, p. 208).

It seems reasonable to suppose that causal explanations—where the causal concepts are construed broadly enough to include probabilistic as well as deterministic relations—are appropriate to non-quantum-mechanical situations. That, at any rate, is the position I shall defend. However, as I shall explain in chapter 9, the question of whether causal concepts, even broadly construed, can correctly be applied in the quantum domain is moot. Even if the answer is negative, however, it still makes sense to speak of a mechanistic conception of scientific explanation, for quantum mechanics is a twentieth-century theory of the physical mechanisms that operate in the world. It is my view that a mechanistic or causal version of the ontic conception provides the most tenable philosophical theory of scientific explanation that is currently available.

HOW THESE CONCEPTIONS ANSWER A FUNDAMENTAL QUESTION

Near the beginning of the first chapter, I mentioned what strikes me as a profound philosophical problem concerning the nature of scientific explanation. If it is true that one of the major aims of science is explanation of natural phenomena, then what sort of knowledge, over and above predictive and descriptive knowledge, is involved in explanation? Put the question this way. Suppose you had achieved the epistemic status of Laplace's demon—that fictitious being who knows all of the laws of nature, who knows the actual state of the universe at one moment (say, now, according to some convenient simultaneity slice), and who can solve any mathematical problem that is in principle amenable to solution. If Laplacian determinism were true, this being would, as Laplace remarked, know in complete detail about every event that ever occurs—past, present, or future. Its descriptive and predictive knowledge would be complete. But how does it stand with respect to explanatory knowledge? Let us consider the answers that arise out of the three general conceptions of scientific explanation we have been discussing.

(1A) According to the *inferential version of the epistemic conception*, which takes explanations to be arguments of various sorts, nothing more is needed. Laplace's demon would have complete explanatory knowledge as well as complete descriptive/predictive knowledge. Still, as we have noted previously, a strange asymmetry appears. Using its knowledge of the laws and present conditions in the universe, the demon could predict any future occurrence; after the predicted event occurs, that very predictive inference would constitute an explanation of the future event. Similarly, knowing the present state of the universe and all of the laws, the demon could retrodict any past event; such retrodictions, however, would not constitute explanations, for we explain events in terms of antecedent facts, not subsequent ones.[17] This would pose no great obstacle to the demon,

---

[17] Hempel (1965, pp. 353–354) tentatively suggests that it may in fact be possible to provide a scientifically acceptable explanation of an event E in terms of circumstances that are not antecedent to it, but either simultaneous with it or subsequent to it. He acknowledges that common sense seems to reject such explanations because subsequent events cannot "bring about" E, but he expresses qualms about the precise meaning of that phrase. We shall try to explicate the concept of producing in chapter 6.

Grünbaum (1973, chap. 9) expresses serious doubts about the asymmetry of explanation.

for it could retrodict some facts still earlier in the history of the universe than the fact-to-be-explained, and it could construct an explanation on the basis of the earlier facts. If explanations are essentially arguments, however, it is hard to understand why this roundabout procedure is required to provide satisfactory explanations. The inferential version of the epistemic conception does not seem to offer an acceptable answer to the question about the distinction between explanation and prediction.

One could, of course, impose restrictions upon the kinds of arguments that may qualify as explanations in ways that would prevent irrelevancies from creeping in (see Fetzer, 1981, pp. 125–126, 137–138) and prescribe appropriate temporal relations between the explanans and the explanandum. The fact that such moves can be made does not, however, answer the fundamental question. The basic issue is to show why, if being an argument is an essential feature of each and every explanation, it is necessary to impose such apparently ad hoc conditions upon the sorts of arguments that have explanatory force. Moreover, there remains the problem, which has been discussed at length previously, of dealing with the explanation of low-probability events.

I am reminded of an old Yiddish riddle: What is green, hangs on a wall, and whistles? The answer is, a herring. But, you reply, a herring is not green. Well, it could be painted green. But, you continue, it does not hang on the wall. Well, you could hang it on the wall. But, you rejoin, herrings do not whistle. Well, you are told with a shrug, two out of three is not too bad.

(1B) There are two distinct ways to construe the *information-theoretic version of the epistemic conception*. If, with Greeno's earlier account, we look only at statistical relevance relations among events and conditions— whether they be directly observable or not—we are open to essentially the same criticism as was just directed against the inferential version. These relevance relations provide a good basis for forming rational expectations about what will take place, but they do not add any kind of knowledge over and above that of the descriptive/predictive variety. When the information-theoretic approach is construed in this way, it is essentially the same theory of scientific explanation as was embodied in the statistical-relevance model, and it has all of the same drawbacks. We found sufficient reasons in chapter 2 to reject that theory precisely because it could not offer the kinds of knowledge about causes and mechanisms that seem indispensable to satisfactory explanation.

There is, however, another way to look at the information-theoretic

---

This view results, I believe, from too strong a commitment to the inferential version of the epistemic conception of scientific explanation.

approach; this way comes closer to matching Sayre's theory (1977). The difference between this view and Greeno's can be put quite simply. In Greeno's approach, we consider the relation between an explanans-partition and an explanandum-partition, and we measure the quantity of information transmitted. On Sayre's view, we must look not only at the amount of information transmitted, but also at the way in which it is transmitted. When the nature of the transmission process is taken into account, we are, as Sayre clearly indicates, dealing with a causal theory. I do not see that there is any basic difference between transmission of information and transmission of causal influence. Information-theoretic concepts have a great many uses—among them, the treatment of problems in the storage and communication of knowledge. They can also be applied more generally to the description of the mechanisms that operate throughout the physical world—mechanisms that are completely independent of our human epistemic concerns. When the information-theoretic account is viewed in this light, it ceases to be (in any important sense) a version of the epistemic conception, but can be taken as a useful variation on the causal conception.

(1C) Among contemporary philosophers who have dealt with scientific explanation, van Fraassen is one of the very few to have explicitly addressed the question of the relation between descriptive/predictive knowledge and explanatory knowledge. In his *erotetic version of the epistemic conception*, he offers an appealing pragmatic answer. He argues, very simply, that all explanatory information is descriptive information—that is, there is no such thing as explanatory knowledge over and above descriptive knowledge. Not all descriptive knowledge, however, is explanatory. What makes an item of descriptive knowledge explanatory is that it can be used to answer a why-question. Why-questions, as we have seen in this study, arise in contexts that are determined by the background knowledge and interests of particular people in their own special situations. Descriptive knowledge that does not answer anyone's why-questions is not explanatory knowledge. This does not mean that it is an inferior sort of knowledge; it means only that no one has happened to request it. "To sum up: no factor is explanatorily relevant unless it is scientifically relevant; and among the scientifically relevant factors, context determines explanatorily relevant ones" (1980, p. 126).

Although van Fraassen offers the following characterization of views other than his own, he seems to agree that, for many contexts, in a rough sense at least, only factors that are causally relevant can fulfill the explanatory function. This is his formulation:

1. Events are enmeshed in a net of causal relations
2. What science describes is that causal net

3. Explanation of why an event happens consists (typically) in an exhibition of salient factors in the part of the causal net formed by the lines 'leading up to' that event
4. Those salient factors mentioned in an explanation constitute (what are ordinarily called) the *cause(s)* of that event.

Interest in causation as such focuses attention on 1 and 2, but interest in explanation requires us to concentrate on 3 and 4. Indeed, from the latter point of view, it is sufficient to guarantee the truth of 1 and 2 by *defining*

the causal net = whatever structure of relations science describes

and leaving to those interested in causation as such the problem of describing that structure in abstract but illuminating ways, if they wish. (1980, p. 124)

Our next three chapters will be devoted to the task of describing causal structures (an enterprise van Fraassen deftly sidesteps, as we have just seen), but not out of an interest in causation as such. Since the exhibition of causal mechanisms is an essential part of scientific explanation on my view, I cannot afford to evade that issue.

In describing the desiderata for explanations, van Fraassen requires that the supposed explanatory facts bear objective causal (and perhaps statistical) relevance relations to the fact-to-be-explained; this is precisely the sort of relevance relation upon which I want to insist. We agree that the drop in the pressure reading on the barometer does not explain the storm because it fails to fulfill that condition. In addition, he makes reference to *salient* factors in a situation which have that status for a particular individual "because of his orientation, his interests, and various other peculiarities in the way he approaches or comes to know the problem—contextual factors" (1980, p. 125). Such salient features are often identified in ordinary situations as 'causes.' To illustrate, he quotes a well-known passage from Norwood Russell Hanson:

> There are as many causes of *x* as there are explanations of *x*. Consider how the cause of death might have been set out by a physician as "multiple haemorrhage," by the barrister as "negligence on the part of the driver," by a carriage-builder as "a defect in the brakeblock construction," by a civic planner as "the presence of tall shrubbery at that turning." (Hanson, 1958, p. 54; quoted by van Fraassen, 1980, p. 125)

It is clear that the civic planner, who is concerned to lay out streets that do not possess unnecessary hazards to their users, is interested in blind intersections and turns; the medical details of traffic fatalities are 'not his department.' The physiological facts are objectively relevant to the occurrence of death, but they are not salient in relation to the concerns of the civic planner. Conversely, the presence of shrubbery at the scene of the accident is not salient to the medical examiner.

Since the relation of salience raises pivotal issues, it will be advisable to examine Hanson's example closely and to discuss salience in some detail. In presenting the S-R basis for a scientific explanation, I emphasized the importance of both the choice of an initial reference class and the specification of the explanandum-partition (van Fraassen's contrast class). To deal with the choice of a reference class, we have to pay special attention to the nature of the x discussed by Hanson, and to deal with the explanandum-partition, we must look closely at the attribute whose occurrence is to be explained.

In considering Hanson's example from the four different standpoints he mentions, we should note immediately that the reference class of the medical examiner is quite different from that of the three other specialists. The medical examiner is concerned to determine why this person died. The reference class is human beings, and the attribute is dying. The explanandum-partition presumably contains two cells: dying and surviving. In the given reference class, the occurrence of multiple hemorrhage is statistically (and causally) relevant to death.

At this point, it may be advisable to comment upon a recurrent issue, namely, the explanatory import of necessary conditions (see van Fraassen, 1980, pp. 114–115). The medical examiner, in attempting to provide an explanation of the fatality, does not include the fact that the decedent was born. Although birth is a necessary antecedent of death, it is not a relevant necessary condition, for it produces no relevant partition—indeed, no partition of any kind—in the reference class. If, however, we ask why a particular individual has paresis, where the reference class is all humans, the answer "Because of latent untreated syphilis" is appropriate. Latent untreated syphilis is a necessary condition for the occurrence of paresis, but it is a relevant necessary condition, since it effects a relevant partition in the reference class. The moral is simple. Relevant antecedent necessary conditions do have explanatory import, if they are not screened off; irrelevant antecedent conditions do not. This point can be applied directly to van Fraassen's examples. The ringing of the alarm clock may explain why a sleeper awakes; the fact that he went to sleep does not.

The barrister, carriage designer, and civic planner are all interested in the occurrence of a carriage accident at a turn. With reasonable plausibility we might, I think, identify the reference class with which they are concerned as the class of passages of carriages through curves, turns, and intersections. Within that class, they want to explain the occurrence of accidents; indeed, they may even ignore the fact, for purposes of explanation, that the accident was a fatal one.

When a barrister argues the case in court, the primary aim may be to find an explanation of this accident, quite apart from the fact that it resulted in a fatality. One wants to know why this particular case of a carriage rounding a turn resulted in an accident. The reference class is the class of events consisting of carriages making turns; the explanandum-partition might consist of: no accident, accident involving no significant property damage or personal injury, accident involving property damage but no personal injury, accident involving personal injury but no property damage, accident involving both personal injury and property damage. The defect in the brake-block construction is certainly relevant, as is the presence of tall shrubbery at the corner. The question is whether, given these other factors, a negligent act on the part of the driver—for example, taking a hazardous curve at too great a speed—is also objectively relevant. If justice is to be done, all significantly relevant factors must be taken into account. If a negligent act is causally relevant, then monetary compensation may be due. The fact that the accident resulted in a fatality, rather than merely minor injuries, will surely have a bearing upon the size of the monetary award, but that legal question is quite different from the question of whether a negligent act constitutes part of the explanation. If, however, the barrister does intend to explain why a fatal accident occurred, the reference class would remain the same, but the explanandum-partition would make separate divisions for death and personal injury. The particular why-question determines which explanation is being sought; when that is determined, the explanation must take all objectively relevant factors into account (unless, of course, the degree of relevance in such practical situations is very slight).

The carriage-builder's concern is quite similar to that of the barrister, and quite different from that of the medical examiner. The question for the carriage-builder is whether, in the reference class of passages of carriages through curves, turns, and intersections, a particular feature of the brake-block construction is relevant to the occurrence of an accident. In this case, the explanandum-partition might contain only two cells—accident and no accident. Like the barrister, the carriage-builder has no license to ignore factors, such as poor visibility or reckless driving, that are objectively relevant to the occurrence that is to be explained.

Similar remarks can be made with regard to the civic planner. The why-question in this case involves the same initial reference class of passages of carriages through turns, curves, and intersections. The planner's explanandum-partition might be different from that of the barrister or the carriage designer; it might consist of: no accident, minor accident, major accident, fatal accident. In any case, the planner wants to know whether, when all other objectively relevant factors are taken into account, the presence of the shrubbery is also relevant.

It seems to me quite possible that the barrister, carriage designer, and civic planner are all concerned with the same explanation. Perhaps there are differences in their contrast classes (to come back to van Fraassen's terminology), but perhaps not. It may be that these three people are all looking for the same set of statistically or causally relevant factors, but one is interested in whether a negligent act is among them, another in whether a design defect is among them, and the third in whether reduced visibility is among them. The fact that one person is more interested than another in one particular relevant factor does not mean that they are either seeking or finding different explanations of the same fact.

It might seem that the interests of the medical examiner are not as different from those of the other three experts as I have made out, for the coroner may need to determine how the fatal injuries were inflicted. If, however, the medical examiner should go on to ask why this person suffered multiple hemorrhage, that is quite a different why-question, and it calls for an altogether different explanation.

The erotetic approach of Bromberger and van Fraassen is valuable for its insistence upon precise articulation of the why-question involved in any request for an explanation. This issue is crucial to the statistical-relevance approach as well, for an appropriate specification of the initial reference class and the explanandum-partition is required to provide the S-R basis. The statistical relevance relations contained within the S-R basis furnish a foundation for the relations of causal relevance to which we appeal for explanatory purposes. As I said in chapter 2, there is no surprise in the fact that different why-questions elicit different explanations. One useful result of the discussions of the pragmatics of explanation is to show how easily a given interrogative sentence, in a given situation, may be used by different people to express different why-questions. This realization translates into the statistical-relevance approach in two ways: first, in the recognition that one and the same event can be a member of many different initial reference classes, and second, in the realization that different explanandum-partitions may yield different explanations. So we can see readily how contextual factors are involved in the correct identification of the why-question that is used to call forth a scientific explanation. It obviously does not follow from all of this that once we have ascertained by reference to pragmatic and contextual factors what explanation is being sought, the explanation itself must embody pragmatic features. Various contextual considerations must be brought to bear to discover what problem the question "Why is the thingumajig all fouled up?" addresses. Nevertheless, when we discover that the question is why the paper is not feeding properly in the photocopy machine, we find that there is an objective and nonpragmatic answer: The feeder was not securely fastened to the machine.

Let us assume, then, that a particular why-question has been specified, and that an answer, "Because A," has been given. Van Fraassen and I agree that certain *objective* relevance relations must obtain between the explanans A and the fact-to-be-explained if that answer is to constitute an acceptable explanation. On van Fraassen's view, however, we must also take account of considerations of 'salience'; these are relations of *pragmatic* relevance. As I understand the notion of salience, an answer to a why-question could fail to be salient to a given questioner for either of two reasons. First, it might convey information that the questioner already has; in van Fraassen's terminology, it is already part of $K(Q)$. This aspect of salience raises no issues of great moment, as far as I can see, though I would be inclined to say that objectively relevant facts, both particular and general, are parts of the explanation whether or not they are already known to us. Second, the answer, "Because A," may contain objectively relevant information that is not already part of the questioner's body of knowledge, but that, nevertheless, fails to be salient simply because the questioner is not interested in A. I think there may be a fundamental difference of opinion between van Fraassen and me on this issue.

Consider a concrete example. Suppose that a member of a congressional committee asks why a particular airplane that took off from Washington National Airport one winter day failed to gain sufficient altitude, with the result that it struck a bridge and crashed into the Potomac River. An official FAA investigator answers that it was because ice had built up on the wing surfaces. The questioner is satisfied. The investigator goes on, nevertheless, to explain that, due to the Bernoulli principle, an airplane derives its lift from wings that have the shape of airfoils. Ice adhering to the leading edge of the wing alters the shape of the surface so that it no longer functions as an effective airfoil, and no longer provides adequate lift. The committee member, uninterested in scientific details, dismisses the latter part of the explanation as irrelevant—in van Fraassen's terms, it is not salient to this person in this context. Hence, as I understand it, van Fraassen would maintain that the information about the Bernoulli principle, the shape of the airfoil, and the lift does not constitute an acceptable part of the answer to the why-question posed by the member of Congress, even though that information does not constitute part of that person's background knowledge.

The fact that the airplane went down in the river is not explained, on my view, unless the causal mechanisms relating the accumulation of ice on the wings to the loss of lift are mentioned. For van Fraassen, in contrast, the buildup of ice on the wings is an adequate explanation if that information fills the gap in the questioner's knowledge that he or she wanted to have filled. The scientific elaboration furnishes explanatory knowledge only if the questioner has scientific curiosity; otherwise, it is not salient. This is the point at which I find van Fraassen's pragmatic view untenable. Nevertheless, he has explicitly addressed the question of the relationship between descriptive knowledge and explanatory knowledge, and he has given an important and intelligible answer.

(2) The *modal conception* offers a straightforward answer to the question about the nature of explanatory knowledge. The descriptive knowledge possessed by Laplace's demon tells us what actually occurs in the world during its entire history. It tells us, for instance, that in every interaction between two bodies, linear momentum *is* conserved. We explain what goes on in such interactions if we can show that linear momentum *must be* conserved. Explanation adds the modal component, which is lacking in purely predictive knowledge. We cannot accept this answer, however, for we have found compelling reasons for rejecting the modal conception.

(3) The *ontic conception* can also provide an answer to this question about explanation. There are several items of knowledge, essential to explanation but not indispensable to prediction, that the demon may lack. First, it is important to distinguish *causal laws* from noncausal laws. There is no indication that Laplace recognized this distinction or required his demon to make it. Second, as we shall see, *causal processes* play a crucial role in explanation. We shall discuss the importance of distinguishing causal processes from pseudo-processes. Again, there is no indication that Laplace was cognizant of this distinction or that he required his demon to notice it. Third, we must make a distinction between *causal interactions* and other kinds of events. Once more, we have no grounds for thinking Laplace was aware of it, or that he required his demon to take it into account. These three causal factors play distinct roles in forming the patterns that we invoke in constructing explanations. Causal processes *propagate* the structure of the physical world and provide the connections among the happenings in the various parts of space-time. Causal interactions *produce* the structure and modifications of structure that we find in the patterns exhibited by the physical world. Causal laws *govern* the causal processes and causal interactions, providing the regularities that characterize the evolution of causal processes and the modifications that result from causal interactions. Causal processes, causal interactions, and causal laws provide the mechanisms by which the world works; to understand *why* certain things happen, we need to see *how* they are produced by these mechanisms.

We can now drop the fiction of Laplace's demon, and the determinism which that mythical being was introduced to explicate. Even if the processes and interactions are stochastic, and the basic laws irreducibly statistical, the foregoing remarks about the relationship between the causal mechanisms and scientific explanation still hold. To understand the world and what goes on in it, we must expose its inner workings. To the extent that causal mechanisms operate, they explain how the world works. If quantum mechanics requires noncausal mechanisms, they also explain what goes on. A detailed knowledge of the mechanisms may not be required for successful prediction; it is indispensable to the attainment of genuine scientific understanding.

CONCLUSIONS

Our rather lengthy discussions of the three leading general conceptions of scientific explanation have left us with two viable alternatives. We have found sufficient reasons, in my opinion, for rejecting the modal conception and the inferential version of the epistemic conception. The information-theoretic version of the epistemic conception survives just in case it devotes enough attention to the causal processes of information transmission (i.e., transmission of causal influence) to qualify for a transfer from the epistemic to the ontic conception. That leaves the erotetic version of the epistemic conception and the ontic conception as significant candidates.

Regardless of our ultimate philosophical views on the nature of scientific explanation, we should recognize that careful attention to the logic of why-questions is an important—possibly indispensable—prolegomenon to the study. Van Fraassen has provided a valuable treatment of this aspect of the theory of scientific explanation. What makes his philosophical theory distinctive is its insistence upon the essentially pragmatic character of explanations. This pragmatic thesis is a crucial component of his view that explanatory knowledge is simply descriptive knowledge that serves particular functions in particular contexts. Since a context involves a body of background knowledge, as well as various goals, interests, and desires, explanation must be considered epistemically relativized.

The ontic—that is, the causal or mechanical—conception has been set in opposition to van Fraassen's pragmatic theory. On this view, I have suggested, explanatory knowledge involves something over and above merely descriptive and/or predictive knowledge, namely, knowledge of underlying mechanisms. This fundamental philosophical difference is intimately associated with the fact that van Fraassen and I come down on opposite sides of the controversy concerning theoretical realism. For van Fraassen, who characterizes his view as "constructive empiricism" (1980, p. 5), scientific theories are judged strictly on the basis of their empirical adequacy, and the acceptance of a theory does not involve a commitment to believe that unobservable objects, to which it apparently makes reference, actually exist. Van Fraassen's constructive empiricism thus precludes appeal, for purposes of explanation, to any 'underlying mechanisms' that involve unobservable objects, events, processes, properties, or relations.[18] It is in this issue, I believe, that the fundamental difference between the pragmatic/epistemic and the causal/mechanical theories of explanation lie. I shall say no more about this issue here, for it constitutes the main topic of chapter 8.

There is one further possibility that should be kept in mind when one compares two divergent explications of what seems prima facie to be a single concept; namely, it may be that there are in fact two distinct explicanda.[19] To the FAA investigator and the member of the congressional committee, the explanation of an airplane crash in terms of ice buildup on the wings may contain all of the information that is *salient* with respect to the aim of preventing future crashes. To an individual who wants to achieve scientific understanding of the performance failures of aircraft, a knowledge of the causal mechanisms is essential. Perhaps, then, there are two different sorts of explanation: explanation that increases our manipulative and predictive abilities, and explanation that increases our scientific understanding of natural phenomena. I suspect that van Fraassen has succeeded admirably in capturing the first of these; it is my hope to be able to shed some light upon the second.

---

[18] Van Fraassen does admit that we sometimes appeal to theories that make mention of unobservables for purposes of explanation (1980, pp. 151–152). These appeals, however, do not imply belief in such theories as true, but only acceptance of them as empirically adequate. To my mind, this does *not* amount to an appeal to the underlying mechanisms as such.

[19] It will be recalled that Carnap's well-known theory of the two concepts of probability (1950, chap. 2) rests upon his view that the preanalytic notion of probability involves two distinct explicanda.

5 | Causal Connections

In PRECEDING chapters, I have made frequent reference to the role of causality in scientific explanation, but I have done nothing to furnish an analysis of the concept of causality or its subsidiary notions. The time has come to focus attention specifically upon this issue, and to see whether we can provide an account of causality adequate for a causal theory of scientific explanation. I shall not attempt to sidestep the fundamental philosophical issues. It seems to me that intellectual integrity demands that we squarely face Hume's incisive critique of causal relations and come to terms with the profound problems he raised.[1]

BASIC PROBLEMS

As a point of departure for the discussion of causality, it is appropriate for us to take a look at the reasons that have led philosophers to develop theories of explanation that do not require causal components. To Aristotle and Laplace it must have seemed evident that scientific explanations are inevitably causal in character. Laplacian determinism is causal determinism, and I know of no reason to suppose that Laplace made any distinction between causal and noncausal laws. In their 1948 paper, Hempel and Oppenheim make the same sort of identification in an offhand manner (Hempel, 1965, p. 250; but see note 6, same page, added in 1964); however, in subsequent writings, Hempel has explicitly renounced this view (e.g., 1965, pp. 352–354).

It might be initially tempting to suppose that all laws of nature are causal laws, and that explanation in terms of laws is ipso facto causal explanation. It is, however, quite easy to find law-statements that do not express causal relations. Many regularities in nature are not direct cause-effect relations. Night follows day, and day follows night; nevertheless, day does not cause night, and night does not cause day. Kepler's laws of planetary motion

---

[1] I find the attempts of Harré and Madden (1975) and Wright (1976) to evade this issue utterly unconvincing. It will be evident from the discussions of this chapter and the next, I trust, that the problems of explicating such concepts as causal connections, causal interactions, and cause-effect relations cannot be set aside as mere philosophical quibbles.

describe the orbits of the planets, but they offer no causal account of these motions.[2] Similarly, the ideal gas law

$$PV = nRT$$

relates pressure (P), volume (V), and temperature (T) for a given sample of gas, and it tells how these quantities vary as functions of one another, but it says nothing whatever about causal relations among them. An increase in pressure might be brought about by moving a piston so as to decrease the volume, or it might be caused by an increase in temperature. The law itself is entirely noncommittal concerning such causal considerations. Each of these regularities—the alternation of night with day; the regular motions of the planets; and the functional relationship among temperature, pressure, and volume of an ideal gas—can be *explained* causally, but they do not *express* causal relations. Moreover, they do not afford causal explanations of the events subsumed under them. For this reason, it seems to me, their value in providing scientific explanations of particular events is, at best, severely limited. These are regularities that need to be explained, but that do not, by themselves, do much in the way of explaining other phenomena.

To untutored common sense, and to many scientists uncorrupted by philosophical training, it is evident that causality plays a central role in scientific explanation. An appropriate answer to an explanation-seeking why-question normally begins with the word "because," and the causal involvements of the answer are usually not hard to find.[3] The concept of causality has, however, been philosophically suspect ever since David Hume's devastating critique, first published in 1739 in his *Treatise of Human Nature*. In the "Abstract" of that work, Hume wrote:

> Here is a billiard ball lying on the table, and another ball moving toward it with rapidity. They strike; the ball which was formerly at rest now acquires a motion. This is as perfect an instance of the relations of cause and effect as any which we know either by sensation or reflection. Let us therefore examine it. It is evident that the two balls touched one another before the motion was communicated, and that there was no interval betwixt the shock and the motion. *Contiguity* in time and place is therefore a requisite circumstance to the operation of all causes. It is evident, likewise, that the motion which was the cause

[2] It might be objected that the alternation of night with day, and perhaps Kepler's 'laws,' do not constitute genuine lawful regularities. This consideration does not really affect the present argument, for there are plenty of regularities, lawful and nonlawful, that do not have explanatory force, but that stand in need of causal explanation.

[3] Indeed, in Italian, there is one word, *perche*, which means both "why" and "because." In interrogative sentences it means "why" and in indicative sentences it means "because." No confusion is engendered as a result of the fact that Italian lacks two distinct words.

is prior to the motion which was the effect. *Priority* in time is, therefore, another requisite circumstance in every cause. But this is not all. Let us try any other balls of the same kind in a like situation, and we shall always find that the impulse of the one produces motion in the other. Here, therefore, is a *third* circumstance, viz., that of *constant conjunction* betwixt the cause and the effect. Every object like the cause produces always some object like the effect. Beyond these three circumstances of contiguity, priority, and constant conjunction I can discover nothing in this cause. (1955, pp. 186–187)

This discussion is, of course, more notable for factors Hume was unable to find than for those he enumerated. In particular, he could not discover any 'necessary connections' relating causes to effects, or any 'hidden powers' by which the cause 'brings about' the effect. This classic account of causation is rightly regarded as a landmark in philosophy.

In an oft-quoted remark that stands at the beginning of a famous 1913 essay, Bertrand Russell warns philosophers about the appeal to causality:

All philosophers, of every school, imagine that causation is one of the fundamental axioms or postulates of science, yet, oddly enough, in advanced sciences such as gravitational astronomy, the word "cause" never occurs. . . . To me it seems that . . . the reason why physics has ceased to look for causes is that, in fact, there are no such things. The law of causality, I believe, like much that passes muster among philosophers, is a relic of a bygone age, surviving, like the monarchy, only because it is erroneously supposed to do no harm. (1929, p. 180)

It is hardly surprising that, in the light of Hume's critique and Russell's resounding condemnation, philosophers with an empiricist bent have been rather wary of the use of causal concepts. By 1927, however, when he wrote *The Analysis of Matter*, Russell recognized that causality plays a fundamental role in physics; in *Human Knowledge*, four of the five postulates he advanced as a basis for all scientific knowledge make explicit reference to causal relations (1948, pp. 487–496). It should be noted, however, that the causal concepts he invokes are *not* the same as the traditional philosophical ones he had rejected earlier.[4] In contemporary physics, causality is a pervasive ingredient (Suppes, 1970, pp. 5–6).

Two Basic Concepts

A standard picture of causality has been around at least since the time of Hume. The general idea is that we have two (or more) distinct events

---

[4] In this latter work (1948), regrettably, Russell felt compelled to relinquish empiricism. I shall attempt to avoid such extreme measures.

that bear some sort of cause-effect relations to one another. There has, of course, been considerable controversy regarding the nature of both the relation and the relata. It has sometimes been maintained, for instance, that facts or propositions (rather than events) are the sorts of entities that can constitute relata. It has long been disputed whether causal relations can be said to obtain among individual events, or whether statements about cause-effect relations implicitly involve assertions about classes of events. The relation itself has sometimes been taken to be that of sufficient condition, sometimes necessary condition, or perhaps a combination of the two.[5] Some authors have even proposed that certain sorts of statistical relations constitute causal relations. This suggestion will be considered at length in chapter 7.

The foregoing characterization obviously fits J. L. Mackie's sophisticated account in terms of INUS conditions—that is, *insufficient* but *nonredundant* parts of *unnecessary* but *sufficient* conditions (1974, p. 62). The idea is this. There are several different causes that might account for the burning down of a house: careless smoking in bed, an electrical short circuit, arson, being struck by lightning. With certain obvious qualifications, each of these may be taken as a sufficient condition for the fire, but none of them can be considered necessary. Moreover, each of the sufficient conditions cited involves a fairly complex combination of conditions, each of which constitutes a nonredundant part of the particular sufficient condition under consideration. The careless smoker, for example, must fall asleep with his cigarette, and it must fall upon something flammable. It must not awaken the smoker by burning him before it falls from his hand. When the smoker does become aware of the fire, it must have progressed beyond the stage at which he can extinguish it. Any one of these necessary components of some complex sufficient condition can, under certain circumstances, qualify as a cause. According to this standard approach, events enjoy the status of fundamental entities, and these entities are 'connected' to one another by cause-effect relations.

It is my conviction that this standard view, in all of its well-known variations, is profoundly mistaken, and that a radically different notion should be developed. I shall not, at this juncture, attempt to mount arguments against the standard conception; some will be suggested in chapter 7. Instead, I shall present a rather different approach for purposes of comparison. I hope that the alternative will stand on its own merits.

There are, I believe, two fundamental causal concepts that need to be explicated, and if that can be achieved, we will be in a position to deal

[5] See (Mackie, 1974) for an excellent historical and systematic survey of the various approaches.

with the problems of causality in general. The two basic concepts are *propagation* and *production*, and both are familiar to common sense. The first of these will be treated in this chapter; the second will be handled in the next chapter. When we say that the blow of a hammer drives a nail, we mean that the impact produces penetration of the nail into the wood. When we say that a horse pulls a cart, we mean that the force exerted by the horse produces the motion of the cart. When we say that lightning ignites a forest, we mean that the electrical discharge produces a fire. When we say that a person's embarrassment was due to a thoughtless remark, we mean that an inappropriate comment produced psychological discomfort. Such examples of causal production occur frequently in everyday contexts.

Causal propagation (or transmission) is equally familiar. Experiences that we had earlier in our lives affect our current behavior. By means of memory, the influence of these past events is transmitted to the present (see Rosen, 1975). A sonic boom makes us aware of the passage of a jet airplane overhead; a disturbance in the air is propagated from the upper atmosphere to our location on the ground. Signals transmitted from a broadcasting station are received by the radio in our home. News or music reaches us because electromagnetic waves are propagated from the transmitter to the receiver. In 1775, some Massachusetts farmers—in initiating the American Revolutionary War—"fired the shot heard 'round the world" (Emerson, 1836). As all of these examples show, what happens at one place and time can have significant influence upon what happens at other places and times. This is possible because causal influence can be propagated through time and space. Although causal production and causal propagation are intimately related to one another, we should, I believe, resist any temptation to try to reduce one to the other.

PROCESSES

One of the fundamental changes that I propose in approaching causality is to take processes rather than events as basic entities. I shall not attempt any rigorous definition of processes; rather, I shall cite examples and make some very informal remarks. The main difference between events and processes is that events are relatively localized in space and time, while processes have much greater temporal duration, and in many cases, much greater spatial extent. In space-time diagrams, events are represented by points, while processes are represented by lines. A baseball colliding with a window would count as an event; the baseball, traveling from the bat to the window, would constitute a process. The activation of a photocell by a pulse of light would be an event; the pulse of light, traveling, perhaps from a distant star, would be a process. A sneeze is an event. The shadow of a cloud moving across the landscape is a process. Although I shall deny that all processes qualify as causal processes, what I mean by a process is similar to what Russell characterized as a causal line:

> A causal line may always be regarded as the persistence of something— a person, a table, a photon, or what not. Throughout a given causal line, there may be constancy of quality, constancy of structure, or a gradual change of either, but not sudden changes of any considerable magnitude. (1948, p. 459)

Among the physically important processes are waves and material objects that persist through time. As I shall use these terms, even a material object at rest will qualify as a process.

Before attempting to develop a theory of causality in which processes, rather than events, are taken as fundamental, I should consider briefly the scientific legitimacy of this approach. In Newtonian mechanics, both spatial extent and temporal duration were absolute quantities. The length of a rigid rod did not depend upon a choice of frame of reference, nor did the duration of a process (such as the length of time between the creation and destruction of a material object). Given two events, in Newtonian mechanics, both the spatial distance and the temporal separation between them were absolute magnitudes. A 'physical thing ontology' was thus appropriate to classical physics. As everyone knows, Einstein's special theory of relativity changed all that. Both the spatial distance and the temporal separation were relativized to frames of reference. The length of a rigid rod and the duration of a temporal process varied from one frame of reference to another. However, as Minkowski showed, there is an invariant quantity—the space-time interval between two events. This quantity is independent of the frame of reference; for any two events, it has the same value in each and every inertial frame of reference. Since there are good reasons for according a fundamental physical status to invariants, it was a natural consequence of the special theory of relativity to regard the world as a collection of events that bear space-time relations to one another. These considerations offer support for what is sometimes called an 'event ontology.'

There is, however, another way (originally developed by A. A. Robb) of approaching the special theory of relativity; it is done entirely with paths of light pulses. At any point in space-time, we can construct the Minkowski light cone—a two-sheeted cone whose surface is generated by the paths of all possible light pulses that converge upon the point (past light cone) and the paths of all possible light pulses that could be emitted from the point (future light cone). When all of the light cones are given, the entire space-time structure of the world is determined (see Winnie, 1977). But light pulses, traveling through space and time, are processes. We can, therefore, base special relativity upon a 'process ontology.' Moreover, this approach can be extended in a natural way to general relativity by taking into account the paths of freely falling material particles; these moving gravitational test particles are also processes (see Grünbaum, 1973, pp. 735–750). It is, consequently, entirely legitimate to approach the space-time structure of the physical world by regarding physical processes as the basic types of physical entities. The theory of relativity does not mandate an 'event ontology.'

Whether one adopts the event-based approach or the process-based approach, causal relations must be accorded a fundamental place in the special theory of relativity. As we have seen, any given event E_O, occurring at a particular space-time point P_O, has an associated double-sheeted light cone. All events that could have a causal influence upon E_O are located in the interior or on the surface of the past light cone, and all events upon which E_O could have any causal influence are located in the interior or on the surface of the future light cone. All such events are *causally connectable* with E_O. Those events that lie on the surface of either sheet of the light cone are said to have a *lightlike separation* from E_O, those that lie within either part of the cone are said to have a *timelike separation* from E_O, and those that are outside of the cone are said to have a *spacelike separation* from E_O. The Minkowski light cone can, with complete propriety, be called "the cone of causal relevance," and the entire space-time structure of special relativity can be developed on the basis of causal concepts (Winnie, 1977).

Special relativity demands that we make a distinction between *causal processes* and *pseudo-processes*. It is a fundamental principle of that theory that light is a *first signal*—that is, no signal can be transmitted at a velocity greater than the velocity of light in a vacuum. There are, however, certain processes that can transpire at arbitrarily high velocities—at velocities vastly exceeding that of light. This fact does not violate the basic relativistic principle, however, for these 'processes' are incapable of serving as signals or of transmitting information. Causal processes are those that are capable of transmitting signals; pseudo-processes are incapable of doing so.

Consider a simple example. Suppose that we have a very large circular building—a sort of super-Astrodome, if you will—with a spotlight mounted at its center. When the light is turned on in the otherwise darkened building, it casts a spot of light upon the wall. If we turn the light on for a brief moment, and then off again, a light pulse travels from the light to the wall. This pulse of light, traveling from the spotlight to the wall, is a paradigm of what we mean by a causal process. Suppose, further, that the spotlight is mounted on a mechanism that makes it rotate. If the light is turned on and set into rotation, the spot of light that it casts upon the wall will move around the outer wall in a highly regular fashion. This 'process'—the moving spot of light—seems to fulfill the conditions Russell used to characterize causal lines, but it is not a causal process. It is a paradigm of what we mean by a pseudo-process.

The basic method for distinguishing causal processes from pseudo-processes is the criterion of mark transmission. A causal process is capable of transmitting a mark; a pseudo-process is not. Consider, first, a pulse of light that travels from the spotlight to the wall. If we place a piece of red glass in its path at any point between the spotlight and the wall, the light pulse, which was white, becomes and remains red until it reaches the wall. A single intervention at one point in the process transforms it in a way that persists from that point on. If we had not intervened, the light pulse would have remained white during its entire journey from the spotlight to the wall. If we do intervene locally at a single place, we can produce a change that is transmitted from the point of intervention onward. We shall say, therefore, that the light pulse constitutes a causal process whether it is modified or not, since in either case it is capable of transmitting a mark. Clearly, light pulses can serve as signals and can transmit messages; remember Paul Revere, "One if by land and two if by sea."

Now, let us consider the spot of light that moves around the wall as the spotlight rotates. There are a number of ways in which we can intervene to change the spot at some point; for example, we can place a red filter at the wall with the result that the spot of light becomes red at that point. But if we make such a modification in the traveling spot, it will not be transmitted beyond the point of interaction. As soon as the light spot moves beyond the point at which the red filter was placed, it will become white again. The mark can be made, but it will not be transmitted. We have a 'process,' which, in the absence of any intervention, consists of a white spot moving regularly along the wall of the building. If we intervene at some point, the 'process' will be modified *at that point*, but it will continue on beyond that point just as if no intervention had occurred. We can, of course, make the spot red at other places if we wish. We can install a red lens in the spotlight, but that does not constitute a *local* intervention at an isolated point in the process itself. We can put red filters at many places along the wall, but that would involve *many* interventions rather than a single one. We could get someone to run around the wall holding a red filter in front of the spot continuously, but that would not constitute an intervention *at a single point* in the 'process.'

This last suggestion brings us back to the subject of velocity. If the spot of light is moving rapidly, no runner could keep up with it, but perhaps a mechanical device could be set up. If, however, the spot moves too rapidly, it would be physically impossible to make the filter travel fast enough to keep pace. No material object, such as the filter, can travel at a velocity greater than that of light, but no such limitation is placed upon the spot on the wall. This can easily be seen as follows. If the spotlight rotates at a fixed rate, then it takes the spot of light a fixed amount of time to make one entire circuit around the wall. If the spotlight rotates once per second, the spot of light will travel around the wall in one second. This fact is independent of the size of the building. We can imagine that without making any change in the spotlight or its rate of rotation, the outer walls are expanded indefinitely. At a certain point, when the radius of the building is a little less than 50,000 kilometers, the spot will be traveling at the speed of light (300,000 km/sec). As the walls are moved still farther out, the velocity of the spot exceeds the speed of light.

To make this point more vivid, consider an actual example that is quite analogous to the rotating spotlight. There is a pulsar in the crab nebula that is about 6,500 light-years away. This pulsar is thought to be a rapidly rotating neutron star that sends out a beam of radiation. When the beam is directed toward us, it sends out radiation that we detect later as a pulse. The pulses arrive at the rate of 30 per second; that is the rate at which the neutron star rotates. Now, imagine a circle drawn with the pulsar at its center, and with a radius equal to the distance from the pulsar to the earth. The electromagnetic radiation from the pulsar (which travels at the speed of light) takes 6,500 years to traverse the radius of this circle, but the 'spot' of radiation sweeps around the circumference of this circle in 1/30th of a second; at that rate, it is traveling at about 4×10^{13} times the speed of light. There is no upper limit on the speed of pseudo-processes.[6]

Another example may help to clarify this distinction. Consider a car traveling along a road on a sunny day. As the car moves at 100 km/hr, its shadow moves along the shoulder at the same speed. The moving car, like any material object, constitutes a causal process; the shadow is a pseudo-process. If the car collides with a stone wall, it will carry the marks of that collision—the dents and scratches—along with it long after the collision has taken place. If, however, only the shadow of the car collides with the stone wall, it will be deformed momentarily, but it will resume its normal shape just as soon as it has passed beyond the wall. Indeed, if the car passes a tall building that cuts it off from the sunlight, the shadow will be obliterated, but it will pop right back into existence as soon as the car has returned to the direct sunlight. If, however, the car is totally

---

[6] (Rothman, 1960) contains a lively discussion of pseudo-processes.

obliterated—say, by an atomic bomb blast—it will not pop back into existence as soon as the blast has subsided.

A given process, whether it be causal or pseudo, has a certain degree of uniformity—we may say, somewhat loosely, that it exhibits a certain structure. The difference between a causal process and a pseudo-process, I am suggesting, is that the causal process transmits its own structure, while the pseudo-process does not. The distinction between processes that do and those that do not transmit their own structures is revealed by the mark criterion. If a process—a causal process—is transmitting its own structure, then it will be capable of transmitting certain modifications in that structure.

In *Human Knowledge*, Russell placed great emphasis upon what he called "causal lines," which he characterized in the following terms:

> A "causal line," as I wish to define the term, is a temporal series of events so related that, given some of them, something can be inferred about the others whatever may be happening elsewhere. A causal line may always be regarded as the persistence of something—a person, table, a photon, or what not. Throughout a given causal line, there may be constancy of quality, constancy of structure, or gradual change in either, but not sudden change of any considerable magnitude. (1948, p. 459)

He then goes on to comment upon the significance of causal lines:

> That there are such more or less self-determined causal processes is in no degree logically necessary, but is, I think, one of the fundamental postulates of science. It is in virtue of the truth of this postulate—if it is true—that we are able to acquire partial knowledge in spite of our enormous ignorance. (Ibid.)

Although Russell seems clearly to intend his causal lines to be what we have called causal processes, his characterization may appear to allow pseudo-processes to qualify as well. Pseudo-processes, such as the spot of light traveling around the wall of our Astrodome, sometimes exhibit great uniformity, and their regular behavior can serve as a basis for inferring the nature of certain parts of the pseudo-process on the basis of observation of other parts. But pseudo-processes are not self-determined; the spot of light is determined by the behavior of the beacon and the beam it sends out. Moreover, the inference from one part of the pseudo-process to another is *not* reliable *regardless of what may be happening elsewhere*, for if the spotlight is switched off or covered with an opaque hood, the inference will go wrong. We may say, therefore, that our observations of the various phenomena going on in the world around us reveal processes that exhibit considerable regularity, but some of these are genuine causal processes and others are pseudo-processes. The causal processes are, as Russell says, self-determined; they transmit their own uniformities of qualitative and structural features. The regularities exhibited by the pseudo-processes, in contrast, are parasitic upon causal regularities exterior to the 'process' itself—in the case of the Astrodome, the behavior of the beacon; in the case of the shadow traveling along the roadside, the behavior of the car and the sun. The ability to transmit a mark is the criterion that distinguishes causal processes from pseudo-processes, for if the modification represented by the mark is propagated, the process is transmitting its own characteristics. Otherwise, the 'process' is not self-determined, and is not independent of what goes on elsewhere.

Although Russell's characterization of causal lines is heuristically useful, it cannot serve as a fundamental criterion for their identification for two reasons. First, it is formulated in terms of our ability to infer the nature of some portions from a knowledge of other portions. We need a criterion that does not rest upon such epistemic notions as knowledge and inference, for the existence of the vast majority of causal processes in the history of the universe is quite independent of human knowers. This aspect of the characterization could, perhaps, be restated nonanthropocentrically in terms of the persistence of objective regularities in the process. The second reason is more serious. To suggest that processes have regularities that persist "whatever may be happening elsewhere" is surely an overstatement. If an extremely massive object should happen to be located in the neighborhood of a light pulse, its path will be significantly altered. If a nuclear blast should occur in the vicinity of a mail truck, the letters that it carries will be totally destroyed. If sunspot activity reaches a high level, radio communication is affected. Notice that, in each of these cases, the factor cited does not occur or exist on the world line of the process in question. In each instance, of course, the disrupting factor initiates processes that intersect with the process in question, but that does not undermine the objection to the claim that causal processes transpire in their self-determined fashion regardless of what is happening elsewhere. A more acceptable statement might be that a causal process would persist even if it were isolated from external causal influences. This formulation, unfortunately, seems at the very least to flirt with circularity, for external causal influences must be transmitted to the locus of the process in question by means of other processes. We shall certainly want to keep clearly in mind the notion that causal processes are not parasitic upon other processes, but it does not seem likely that this rough idea could be transformed into a useful basic criterion.

It has often been suggested that the principal characteristic of causal processes is that they transmit energy. While I believe it is true that all and only causal processes transmit energy, there is, I think, a fundamental problem involved in employing this fact as a basic criterion—namely, we must have some basis for distinguishing situations in which energy is transmitted from those in which it merely appears in some regular fashion. The difficulty is easily seen in the 'Astrodome' example. As a light pulse travels from the central spotlight to the wall, it carries radiant energy; this energy is present in the various stages of the process as the pulse travels from the lamp to the wall. As the spot of light travels around the wall, energy appears at the places occupied by the spot, but we do not want to say that this energy is transmitted. The problem is to distinguish the cases in which a given bundle of energy is transmitted through a process from those in which different bundles of energy are appearing in some regular fashion. The key to this distinction is, I believe, the mark method. Just as the detective makes his mark on the murder weapon for purposes of later identification, so also do we make marks in processes so that the energy present at one space-time locale can be identified when it appears at other times and places.

A causal process is one that is self-determined and not parasitic upon other causal influences. A causal process is one that transmits energy, as well as information and causal influence. The fundamental criterion for distinguishing self-determined energy transmitting processes from pseudo-processes is the capability of such processes of transmitting marks. In the next section, we shall deal with the concept of transmission in greater detail.

Our main concern with causal processes is their role in the propagation of causal influences; radio broadcasting presents a clear example. The transmitting station sends a carrier wave that has a certain structure—characterized by amplitude and frequency, among other things—and modifications of this wave, in the form of modulations of amplitude (AM) or frequency (FM), are imposed for the purpose of broadcasting. Processes that transmit their own structures are capable of transmitting marks, signals, information, energy, and causal influence. Such processes are the means by which causal influence is propagated in our world. Causal influences, transmitted by radio, may set your foot to tapping, or induce someone to purchase a different brand of soap, or point a television camera aboard a spacecraft toward the rings of Saturn. A causal influence transmitted by a flying arrow can pierce an apple on the head of William Tell's son. A causal influence transmitted by sound waves can make your dog come running. A causal influence transmitted by ink marks on a piece of paper can gladden one's day or break someone's heart.

It is evident, I think, that the propagation or transmission of causal influence from one place and time to another must play a fundamental role in the causal structure of the world. As I shall argue next, causal processes constitute precisely the causal connections that Hume sought, but was unable to find.

THE 'At-At' THEORY OF CAUSAL PROPAGATION

In the preceding section, I invoked Reichenbach's mark criterion to make the crucial distinction between causal processes and pseudo-processes. Causal processes are distinguished from pseudo-processes in terms of their ability to transmit marks. In order to qualify as causal, a process need not actually be transmitting a mark; the requirement is that it be capable of doing so.

When we characterize causal processes partly in terms of their ability to transmit marks, we must deal explicitly with the question of whether we have violated the kinds of strictures Hume so emphatically expounded. He warned against the uncritical use of such concepts as 'power' and 'necessary connection.' Is not the *ability to transmit* a mark an example of just such a mysterious power? Kenneth Sayre expressed his misgivings on this score when, after acknowledging the distinction between causal interactions and causal processes, he wrote:

> The causal process, continuous though it may be, is made up of individual events related to others in a causal nexus. . . . it is by virtue of the relations among the members of causal series that we are enabled to make the inferences by which causal processes are characterized. . . . if we do not have an adequate conception of the relatedness between individual members in a causal series, there is a sense in which our conception of the causal process itself remains deficient. (1977, p. 206)

The 'at-at' theory of causal transmission is an attempt to remedy this deficiency.

Does this remedy illicitly invoke the sort of concept Hume proscribed? I think not. Ability to transmit a mark can be viewed as a particularly important species of constant conjunction—the sort of thing Hume recognized as observable and admissible. It is a matter of performing certain kinds of experiments. If we place a red filter in a light beam near its source, we can observe that the mark—redness—appears at all places to which the beam is subsequently propagated. This fact can be verified by experiments as often as we wish to perform them. If, contrariwise (returning to our Astrodome example of the preceding section), we make the spot on the wall red by placing a filter in the beam at one point just before the light strikes the wall (or by any other means we may devise), we will see that the mark—redness—is not present at all other places in which the moving spot subsequently appears on the wall. This, too, can be verified by repeated experimentation. Such facts are straightforwardly observable.

The question can still be reformulated. What do we mean when we speak of *transmission*? How does the process *make* the mark appear elsewhere within it? There is, I believe, an astonishingly simple answer. The transmission of a mark from point A in a causal process to point B in the same process *is* the fact that it appears at each point between A and B *without further interactions*. If A is the point at which the red filter is inserted into the beam going from the spotlight to the wall, and B is the point at which the beam strikes the wall, then only the interaction at A is required. If we place a white card in the beam at any point between A and B, we will find the beam red at that point.

The basic thesis about mark transmission can now be stated (in a principle I shall designate MT for "mark transmission") as follows:

> MT: *Let P be a process that, in the absence of interactions with other processes, would remain uniform with respect to a characteristic Q, which it would manifest consistently over an interval that includes both of the space-time points A and B (A ≠ B). Then, a* mark *(consisting of a modification of Q into Q'), which has been introduced into process P by means of a single local interaction at point A, is* transmitted *to point B if P manifests the modification Q' at B and at all stages of the process between A and B without additional interventions.*

This principle is clearly counterfactual, for it states explicitly that the process P would have continued to manifest the characteristic Q if the specific marking interaction had not occurred. This subjunctive formulation is required, I believe, to overcome an objection posed by Nancy Cartwright (in conversation) to previous formulations. The problem is this. Suppose our rotating beacon is casting a white spot that moves around the wall, and that we mark the spot by interposing a red filter at the wall. Suppose further, however, that a red lens has been installed in the beacon just a tiny fraction of a second earlier, so that the spot on the wall becomes red at the moment we mark it with our red filter, but it remains red from that point on because of the red lens. Under these circumstances, were it not for the counterfactual condition, it would appear that we had satisfied the requirement formulated in MT, for we have marked the spot by a single interaction at point A, and the spot remains red from that point on to any other point B we care to designate, without any additional interactions. As we have just mentioned, the installation of the red lens on the spotlight does not constitute a marking of the spot on the wall. The counterfactual stipulation given in the first sentence of MT blocks situations, of the sort mentioned by Cartwright, in which we would most certainly want to deny that any mark transmission occurred via the spot moving around the wall. In this case, the moving spot would have turned red because of the lens even if no marking interaction had occurred locally at the wall.

A serious misgiving arises from the use of counterfactual formulations to characterize the distinction between causal processes and pseudo-processes; it concerns the question of objectivity. The distinction is fully objective. It is a matter of fact that a light pulse constitutes a causal process, while a shadow is a pseudo-process. Philosophers have often maintained, however, that counterfactual conditionals involve unavoidably pragmatic aspects. Consider the famous example about Verdi and Bizet. One person might say, "If Verdi had been a compatriot of Bizet, then Verdi would have been French," whereas another might maintain, "If Bizet had been a compatriot of Verdi, then Bizet would have been Italian." These two statements seem incompatible with one another. Their antecedents are logically equivalent; if, however, we accept both conditionals, we wind up with the conclusion that Verdi would be French, that Bizet would be Italian, and they would still not be compatriots. Yet both statements can be true. The first person could be making an unstated presupposition that the nationality of Bizet is fixed in this context, while the second presupposes that the nationality of Verdi is fixed. What remains fixed and what is subject to change—which are established by pragmatic features of the context in which the counterfactual is uttered—determine whether a counterfactual is true or false. It is concluded that counterfactual conditional statements do not express objective facts of nature; indeed, van Fraassen (1980, p. 118) goes so far as to assert that science contains no counterfactuals. If that sweeping claim were true (which I seriously doubt),[7] the foregoing criterion MT would be in serious trouble.

Although MT involves an explicit counterfactual, I do not believe that the foregoing difficulty is insurmountable. Science has a direct way of dealing with the kinds of counterfactual assertions we require, namely, the experimental approach. In a well-designed controlled experiment, the experimenter determines which conditions are to be fixed for purposes of the experiment and which allowed to vary. The result of the experiment es-

---

[7] For example, our discussion of the Minkowski light cone made reference to paths of possible light rays; such a path is one that would be taken by a light pulse if it were emitted from a given space-time point in a given direction. Special relativity seems to be permeated with reference to possible light rays and possible causal connections, and these involve counterfactuals quite directly. See (Salmon, 1976) for further elaboration of this issue, not only with respect to special relativity but also in relation to other domains of physics. A strong case can be made, I believe, for the thesis that counterfactuals are scientifically indispensable.

tablishes some counterfactual statements as true and others as false under well-specified conditions. Consider the kinds of cases that concern us; such counterfactuals can readily be tested experimentally. Suppose we want to see whether the beam traveling from the spotlight to the wall is capable of transmitting the red mark. We set up the following experiment. The light will be turned on and off one hundred times. At a point midway between the spotlight and the wall, we station an experimenter with a random number generator. Without communicating with the experimenter who turns the light on and off, this second experimenter uses his device to make a random selection of fifty trials in which he will make a mark and fifty in which he will not. If all and only the fifty instances in which the marking interaction occurs are those in which the spot on the wall is red, as well as all the intervening stages in the process, then we may conclude with reasonable certainty that the fifty cases in which the beam was red subsequent to the marking interaction are cases in which the beam would not have been red if the marking interaction had not occurred. On any satisfactory analysis of counterfactuals, it seems to me, we would be justified in drawing such a conclusion. It should be carefully noted that I am *not* offering the foregoing experimental procedure as an analysis of counterfactuals; it is, indeed, a result that we should expect any analysis to yield.

A similar experimental approach could obviously be taken with respect to the spot traversing the wall. We design an experiment in which the beacon will rotate one hundred times, and each traversal will be taken as a separate process. We station an experimenter with a random number generator at the wall. Without communicating with the experimenter operating the beacon, the one at the wall makes a random selection of fifty trials in which to make the mark and fifty in which to refrain. If it turns out that some or all of the trials in which no interaction occurs are, nevertheless, cases in which the spot on the wall turns red as it passes the second experimenter, then we know that we are *not* dealing with cases in which the process will not turn from white to red if no interaction occurs. Hence, if in some cases the spot turns red and remains red after the mark is imposed, we know we are not entitled to conclude that the mark has actually been transmitted.

The account of mark transmission embodied in principle MT—which is the proposed foundation for the concept of propagation of causal influence—may seem too trivial to be taken seriously. I believe such a judgment would be mistaken. My reason lies in the close parallel that can be drawn between the foregoing solution to the problem of mark transmission and the solution of an ancient philosophical puzzle.

About twenty-five hundred years ago, Zeno of Elea enunciated some famous paradoxes of motion, including the well-known paradox of the flying arrow. This paradox was not adequately resolved until the early part of the twentieth century. To establish an intimate connection between this problem and our problem of causal transmission, two observations are in order. First, a physical object (such as the arrow) moving from one place to another constitutes a causal process, as can be demonstrated easily by application of the mark method—for example, initials carved on the shaft of the arrow before it is shot are present on the shaft after it hits its target. And there can be no doubt that the arrow propagates causal influence. The hunter kills his prey by releasing the appropriately aimed arrow; the flying arrow constitutes the causal connection between the cause (release of the arrow from the bow under tension) and the effect (death of a deer). Second, Zeno's paradoxes were designed to prove the absurdity not only of motion, but also of every kind of process or change. Henri Bergson expressed this point eloquently in his discussion of what he called "the cinematographic view of becoming." He invites us to consider any process, such as the motion of a regiment of soldiers passing in review. We can take many snapshots—static views—of different stages of the process, but, he argues, we cannot really capture the movement in this way, for,

> every attempt to reconstitute change out of states implies the absurd proposition, that movement is made out of immobilities.
>
> Philosophy perceived this as soon as it opened its eyes. The arguments of Zeno of Elea, although formulated with a very different intention, have no other meaning.
>
> Take the flying arrow.
>
> (1911, p. 308; quoted in Salmon, 1970a, p. 63)

Let us have a look at this paradox. At any given instant, Zeno seems to have argued, the arrow is where it is, occupying a portion of space equal to itself. During the instant it cannot move, for that would require the instant to have parts, and an instant is *by definition* a minimal and indivisible element of time. If the arrow did move during the instant, it would have to be in one place at one part of the instant and in a different place at another part of the instant. Moreover, for the arrow to move during the instant would require that during that instant it must occupy a space larger than itself, for otherwise it has no room to move. As Russell said:

> It is never moving, but in some miraculous way the change of position has to occur *between* the instants, that is to say, not at any time whatever. This is what M. Bergson calls the cinematographic representation of reality. The more the difficulty is meditated, the more real it becomes.
> (1929, p. 187; quoted in Salmon, 1970a, p. 51)

There is a strong temptation to respond to this paradox by pointing out that the differential calculus provides us with a perfectly meaningful definition of instantaneous velocity, and that this quantity *can* assume values other than zero. Velocity is change of position with respect to time, and the derivative dx/dt furnishes an expression that can be evaluated for particular values of t. Thus an arrow can be at rest at a given moment—that is, dx/dt may equal 0 for that particular value of t. Or it can be in motion at a given moment—that is, dx/dt might be 100 km/hr for another particular value of t. Once we recognize this elementary result of the infinitesimal calculus, it is often suggested, the paradox of the flying arrow vanishes.

This appealing attempt to resolve the paradox is, however, unsatisfactory, as Russell clearly realized. The problem lies in the definition of the derivative; dx/dt is defined as the limit as Δt approaches 0 of $\Delta x/\Delta t$, where Δt represents a nonzero interval of time and Δx may be a nonzero spatial distance. In other words, instantaneous velocity is defined as the limit, as we take decreasing time intervals, of the noninstantaneous average velocity with which the object traverses what is—in the case of nonzero values— a nonzero stretch of space. Thus in the definition of instantaneous velocity, we employ the concept of noninstantaneous velocity, which is precisely the problematic concept from which the paradox arises. To put the same point in a different way, the concept of instantaneous velocity does not genuinely characterize the motion of an object at an isolated instant all by itself, for the very definition of instantaneous velocity makes reference to neighboring instants of time and neighboring points of space. To find an adequate resolution of the flying arrow paradox, we must go deeper.

To describe the motion of a body, we express the relation between its position and the moments of time with which we are concerned by means of a mathematical function; for example, the equation of motion of a freely falling particle near the surface of the earth is

$$x = f(t) = 1/2gt^2 \tag{1}$$

where $g = 9.8$ m/sec^2. We can therefore say that this equation furnishes a function $f(t)$ that relates the position x to the time t. But what is a mathematical function? It is a set of pairs of numbers; for each admissible value of t, there is an associated value of x. To say that an object moves in accordance with equation (1) is simply to say that *at* any given moment t it is *at* point x, where the correspondence between the values of t and of x is given by the set of pairs of numbers that constitute the function represented by equation (1). To move from point A to point B is simply to be *at* the appropriate point of space *at* the appropriate moment of time— no more, no less. The resulting theory is therefore known as "the 'at-at' theory of motion.'' To the best of my knowledge, it was first clearly formulated and applied to the arrow paradox by Russell.

According to the 'at-at' theory, to move from A to B is simply to occupy the intervening points at the intervening instants. It consists in being *at* particular points of space *at* corresponding moments. There is no *additional* question as to how the arrow *gets from* point A *to* point B; the answer has already been given—by being at the intervening points at the intervening moments. The answer is emphatically *not* that it gets from A to B by zipping through the intermediate points at high speed. Moreover, there is no additional question about how the arrow gets from one intervening point to another—the answer is the same, namely, by being at the points between them at the corresponding moments. And clearly, there can be no question about how the arrow gets from one point to the next, for in a continuum there is no next point. I am convinced that Zeno's arrow paradox is a profound problem concerning the nature of change and motion, and that its resolution by Russell in terms of the 'at-at' theory of motion represents a distinctly nontrivial achievement.[8] The fact that this solution can—if I am right—be extended in a direct fashion to provide a resolution of the problem of mark transmission is an additional laurel.

The 'at-at' theory of mark transmission provides, I believe, an acceptable basis for the mark method, which can in turn serve as the means to distinguish causal processes from pseudo-processes. The world contains a great many types of causal processes—transmission of light waves, motion of material objects, transmissions of sound waves, persistence of crystalline structure, and so forth. Processes of any of these types may occur without having any mark imposed. In such instances, the processes still qualify as causal. *Ability* to transmit a mark is the criterion of causal processes; processes that are *actually* unmarked may be causal. Unmarked processes exhibit some sort of persistent structure, as Russell pointed out in his characterization of causal lines; in such cases, we say that the structure is transmitted within the causal process. Pseudo-processes may also exhibit persistent structure; in these cases, we maintain that the structure is *not transmitted* by means of the 'process' itself, but by some other external agency.

The basis for saying that the regularity in the causal process is transmitted via the process itself lies in the ability of the causal process to transmit a modification in its structure—a mark—resulting from an interaction. Consider a brief pulse of white light; it consists of a collection of photons of

---

[8] Zeno's arrow paradox and its resolution by means of the 'at-at' theory of motion are discussed in (Salmon, 1975; 2nd ed., 1980, chap. 2). Relevant writings by Bergson and Russell are reprinted in (Salmon, 1970a); the introduction to this anthology also contains a discussion of the arrow paradox.

various frequencies, and if it is not polarized, the waves will have various spatial orientations. If we place a red filter in the path of this pulse, it will absorb all photons with frequencies falling outside of the red range, allowing only those within that range to pass. The resulting pulse has its structure modified in a rather precisely specifiable way, and the fact that this modification persists is precisely what we mean by claiming that the mark is transmitted. The counterfactual clause in our principle MT is designed to rule out structural changes brought about by anything other than the marking interaction. The light pulse could, alternatively, have been passed through a polarizer. The resulting pulse would consist of photons having a specified spatial orientation instead of the miscellaneous assortment of orientations it contained before encountering the polarizer. The principle of structure transmission (ST) may be formulated as follows:

ST: *If a process is capable of transmitting changes in structure due to marking interactions, then that process can be said to transmit its own structure.*

The fact that a process does not transmit a particular type of mark, however, does not mean that it is not a causal process. A ball of putty constitutes a causal process, and one kind of mark it will transmit is a change in shape imposed by indenting it with the thumb. However, a hard rubber ball is equally a causal process, but it will not transmit the same sort of mark, because of its elastic properties. The fact that a particular sort of structural modification does not persist, because of some inherent tendency of the process to resume its earlier structure, does not mean it is not transmitting its own structure; it means only that we have not found the appropriate sort of mark for that kind of process. A hard rubber ball can be marked by painting a spot on it, and that mark will persist for a while.

Marking methods are sometimes used in practice for the identification of causal processes. As fans of Perry Mason are aware, Lieutenant Tragg always placed 'his mark' upon the murder weapon found at the scene of the crime in order to be able to identify it later at the trial of the suspect. Radioactive tracers are used in the investigation of physiological processes—for example, to determine the course taken by a particular substance ingested by a subject. Malodorous substances are added to natural gas used for heating and cooking in order to ascertain the presence of leaks; in fact, one large chemical manufacturer published full-page color advertisements in scientific magazines for its product "La Stink."

One of the main reasons for devoting our attention to causal processes is to show how they can transmit causal influence. In the case of causal processes used to transmit signals, the point is obvious. Paul Revere was caused to start out on his famous night ride by a light signal sent from the tower of the Old North Church. A drug, placed surreptitiously in a drink, can cause a person to lose consciousness because it retains its chemical structure as it is ingested, absorbed, and circulated through the body of the victim. A loud sound can produce a painful sensation in the ears because the disturbance of the air is transmitted from the origin to the hearer. Radio signals sent to orbiting satellites can activate devices aboard because the wave retains its form as it travels from earth through space. The principle of propagation of causal influence (PCI) may be formulated as follows:

PCI: *A process that transmits its own structure is capable of propagating a causal influence from one space-time locale to another.*

The propagation of causal influence by means of causal processes *constitutes*, I believe, the mysterious connection between cause and effect which Hume sought.

In offering the 'at-at' theory of mark transmission as a basis for distinguishing causal processes from pseudo-processes, we have furnished an account of the transmission of information and propagation of causal influence without appealing to any of the 'secret powers' which Hume's account of causation soundly proscribed. With this account we see that the mysterious connection between causes and effects is not very mysterious after all.

Our task is by no means finished, however, for this account of transmission of marks and propagation of causal influence has used the unanalyzed notion of a causal interaction that produces a mark. Unless a satisfactory account of causal interaction and mark production can be provided, our theory of causality will contain a severe lacuna. We will attempt to fill that gap in the next chapter. Nevertheless, we have made significant progress in explicating the fundamental concept, introduced at the beginning of the chapter, of *causal propagation* (or *transmission*).

This chapter is entitled "Causal Connections," but little has actually been said about the way in which causal processes provide the connection between cause and effect. Nevertheless, in many common-sense situations, we talk about causal relations between pairs of spatiotemporally separated events. We might say, for instance, that turning the key causes the car to start. In this context we assume, of course, that the electrical circuitry is intact, that the various parts are in good working order, that there is gasoline in the tank, and so forth, but I think we can make sense of a cause-effect relation only if we can provide a *causal connection* between the cause and the effect. This involves tracing out the causal processes that lead from the turning of the key and the closing of an electrical circuit to various occurrences that eventuate in the turning over of the engine and the ignition of fuel in the cylinders. We say, for another example, that a tap on the knee causes the foot to jerk. Again, we believe that there are neural impulses traveling from the place at which the tap occurred to the muscles that control the movement of the foot, and processes in those muscles that lead to movement of the foot itself. The genetic relationship between parents and offspring provides a further example. In this case, the molecular biologist refers to the actual process of information transmission via the DNA molecule employing the 'genetic code.'

In each of these situations, we analyze the cause-effect relations in terms of three components—an event that constitutes the cause, another event that constitutes the effect, and a causal process that connects the two events. In some cases, such as the starting of the car, there are many intermediate events, but in such cases, the successive intermediate events are connected to one another by spatiotemporally continuous causal processes. A splendid example of multiple causal connections was provided by David Kaplan. Several years ago, he paid a visit to Tucson, just after completing a boat trip through the Grand Canyon with his family. The best time to take such a trip, he remarked, is when it is very hot in Phoenix. What is the causal connection to the weather in Phoenix, which is about 200 miles away? At such times, the air-conditioners in Phoenix are used more heavily, which places a greater load on the generators at the Glen Canyon Dam (above the Grand Canyon). Under these circumstances, more water is allowed to pass through the turbines to meet the increased demand for power, which produces a greater flow of water down the Colorado River. This results in a more exciting ride through the rapids in the Canyon.

In the next chapter, we shall consider events—especially causal interactions—more explicitly. It will then be easier to see how causal processes constitute precisely the physical connections between causes and effects that Hume sought—what he called "the cement of the universe." These causal connections will play a vital role in our account of scientific explanation.

It is tempting, of course, to try to reduce causal processes to chains of events; indeed, people frequently speak of causal chains. Such talk can be seriously misleading if it is taken to mean that causal processes are composed of discrete events that are serially ordered so that any given event has an immediate successor. If, however, the continuous character of causal processes is kept clearly in mind, I would not argue that it is philosophically incorrect to regard processes as collections of events. At the same time, it does seem heuristically disadvantageous to do so, for this practice seems almost inevitably to lead to the puzzle (articulated by Sayre in the quotation given previously) of how these events, which make up a given process, are causally related to one another. The point of the 'at-at' theory, it seems to me, is to show that no such question about the causal relations among the constituents of the process need arise—for the same reason that, aside from occupying intermediate positions at the appropriate times, there is no further question about how the flying arrow gets from one place to another. With the aid of the 'at-at' theory, we have a complete answer to Hume's penetrating question about the nature of causal connections. For this heuristic reason, then, I consider it advisable to resist the temptation always to return to formulations in terms of events.

6 | Causal Forks and Common Causes

THERE is a familiar pattern of causal reasoning that we all use every day, usually without being consciously aware of it. Confronted with what appears to be an improbable coincidence, we seek a common cause. If the common cause can be found, it is invoked to explain the coincidence.

CONJUNCTIVE FORKS

Suppose, for example, that several members of a traveling theatrical company who have spent a pleasant day in the country together become violently ill that evening. We infer that it was probably due to a common meal of which they all partook. When we find that their lunch included some poisonous mushrooms that they had gathered and cooked, we have the explanation. There is a certain small chance that a particular actor or actress will, on any given evening, suffer severe gastrointestinal distress—the probability need not, of course, be the same for each person. If the illnesses were statistically independent of one another, then the probability of all of the picnickers becoming ill on the same night would be equal to the product of all of these individual small probabilities. Even though a chance coincidence of this sort is possible, it is too improbable to be accepted as such. The evidence for a common cause is compelling.

Although reasoning of this type seems simple and straightforward, philosophers have paid surprisingly little explicit attention to it. Hans Reichenbach is the outstanding exception. In his posthumous book, *The Direction of Time* (1956), he enunciated *the principle of the common cause*, and he attempted to explicate the principle in terms of a statistical structure that he called a *conjunctive fork*. The principle of the common cause states, roughly, that when apparent coincidences occur that are too improbable to be attributed to chance, they can be explained by reference to a common causal antecedent. This principle is by no means trivial or vacuous. Among other things, it denies that such apparent coincidences are to be explained teleologically in terms of subsequent common effects. If the aforementioned theatrical troupe had been scheduled to put on a performance that evening, it would in all likelihood have been canceled. This common effect would not, however, explain the coincidence. We shall have to consider why this is so later in this chapter.

Other examples, from everyday life and from science, are easy to find. If, for instance, two students in a class turn in identical term papers, and if we can rule out the possibility that either copied directly from the other, then we search for a common cause—for example, a paper in a fraternity file from which both of them copied independently of each other.

A recent astronomical discovery, which has considerable scientific significance, furnishes a particularly fine example. The twin quasars 0975 + 561 A and B are separated by an angular width of 5.7 seconds of arc. Two quasars in such apparent proximity would be a rather improbable occurrence given simply the observed distribution of quasars. Examination of their spectra indicates equal red shifts, and hence, equal distances. Thus these objects are close together in space, as well as appearing close together as seen from earth. Moreover, close examination of their spectra reveals a striking similarity—indeed, they are indistinguishable. This situation is in sharp contrast to the relations between the spectra of any two quasars picked at random. Astronomers immediately recognized the need to explain this astonishing coincidence in terms of some sort of common cause. One hypothesis that was entertained quite early was that twin quasars had somehow (no one had the slightest idea how this could happen in reality) developed from a common ancestor. Another hypothesis was the gravitational lens effect—that is, there are not in fact two distinct quasars, but the two images were produced from a single body by the gravitational bending of the light by an intervening massive object. This result might be produced by a black hole, it was theorized, or by a very large elliptical galaxy. Further observation, under fortuitously excellent viewing conditions, has subsequently revealed the presence of a galaxy that would be adequate to produce the gravitational splitting of the image. This explanation is now, to the best of my knowledge, accepted by virtually all of the experts (Chaffee, 1980).

In an attempt to characterize the structure of such examples of common causes, Reichenbach (1956, sec. 19) introduced the notion of a *conjunctive fork*, defined in terms of the following four conditions:[1]

$$P(A.B|C) = P(A|C) \times P(B|C) \tag{1}$$
$$P(A.B|\bar{C}) = P(A|\bar{C}) \times P(B|\bar{C}) \tag{2}$$

[1] The probabilities that appear in these formulas must, I think, be construed as physical probabilities—that is, as frequencies or propensities. In Chapter 7 I shall give my reasons for rejecting the propensity interpretation; hence I construe them as frequencies. Thus I take the variables $A, B, C, \ldots$ which appear in the probability expressions to range over classes.

$$P(A|C) > P(A|\bar{C}) \tag{3}$$
$$P(B|C) > P(B|\bar{C}) \tag{4}$$

For reasons that will be made clear in this chapter, we shall stipulate that none of the probabilities occurring in these relations is equal to zero or one. Although it is not immediately obvious, conditions (1)–(4) entail

$$P(A.B) > P(A) \times P(B) \tag{5}$$

(see Reichenbach, 1956, pp. 160–161).[2] These relations apply quite straightforwardly in concrete situations. Given two effects, A and B, that occur together more frequently than they would if they were statistically independent of one another, there is some prior event C, which is a cause of A and is also a cause of B, that explains the lack of independence between A and B. In the case of plagiarism, the cause C is the presence of the term paper in the file to which both students had access. In the case of simultaneous illness, the cause C is the common meal that included the poisonous mushrooms. In the case of the twin quasar image, the cause C is the emission of radiation in two slightly different directions by a single luminous body.

To say of two events, X and Y, that they occurred independently of one another means that they occur together with a probability equal to the product of the probabilities of their separate occurrences; this is,

$$P(X.Y) = P(X) \times P(Y). \tag{6}$$

Thus in the examples we have considered, as relation (5) states, the two effects A and B are not independent. However, given the occurrence of

---

[2] Reichenbach's proof goes as follows. By the theorem on total probability, we may write:

$$P(A.B) = P(C) \times P(A.B|C) + P(\bar{C}) \times P(A.B|\bar{C}) \tag{a}$$
$$P(A) = P(C) \times P(A|C) + P(\bar{C}) \times P(A|\bar{C}) \tag{b}$$
$$P(B) = P(C) \times P(B|C) + P(\bar{C}) \times P(B|\bar{C}) \tag{c}$$

By virtue of equations (1) and (2), (a) can be rewritten:

$$P(A.B) = P(C) \times P(A|C) \times P(B|C) + P(\bar{C}) \times P(A|\bar{C}) \times P(B|\bar{C}) \tag{d}$$

Now (b), (c), and (d) can be combined to yield:

$$P(A.B) - P(A) \times P(B) =$$
$$P(C) \times P(A|C) \times P(B|C) + P(\bar{C}) \times P(A|\bar{C}) \times P(B|\bar{C}) -$$
$$[P(C) \times P(A|C) + P(\bar{C}) \times P(A|\bar{C})] \times [P(C) \times P(B|C) + P(\bar{C}) \times P(B|\bar{C})] \tag{e}$$

Recalling that $P(\bar{C}) = 1 - P(C)$, we can, by elementary algebraic operations, transform the right-hand side of (e) into the following form:

$$P(C) \times [1 - P(C)] \times [P(A|C) - P(A|\bar{C})] \times [P(B|C) - P(B|\bar{C})]. \tag{f}$$

Assuming that $0 < P(C) < 1$, we see immediately from formulas (3) and (4) that (f) is positive. That result concludes the proof that inequality (5) follows from formulas (1)–(4).

the common cause C, A and B do occur independently, as the relationship among the conditional probabilities in equation (1) shows. Thus in the case of illness, the fact that the probability of two individuals being ill at the same time is greater than the product of the probabilities of their individual illnesses is explained by the common meal. In this example, we are assuming that the fact that one person is afflicted does not have any direct causal influence upon the illness of the other.[3] Moreover, let us assume for the sake of simplicity that in this situation, there are no other potential common causes of severe gastrointestinal illness.[4] Then, in the absence of the common cause C—that is, when $\bar{C}$ obtains—A and B are also independent of one another, as the relationship among the conditional probabilities in equation (2) states. Relations (3) and (4) simply assert that C is a positive cause of A and B, since the probability of each is greater in the presence of C than in the absence of C.

There is another useful way to look at equations (1) and (2). Recalling that, according to the multiplication theorem,

$$P(A.B|C) = P(A|C) \times P(B|A.C), \tag{7}$$

we see that, provided $P(A|C) \neq 0$, equation (1) entails

$$P(B|C) = P(B|A.C). \tag{8}$$

In Reichenbach's terminology, this says that C screens off A from B. A similar argument shows that $\bar{C}$ screens off B from A. To screen off *means* to make statistically irrelevant. Thus, according to equation (1), the common cause C makes each of the two effects A and B statistically irrelevant to one another. By applying the same argument to equation (2), we can easily see that it entails that the absence of the common cause also screens off A from B.

To make quite clear the nature of the conjunctive fork, I should like to use an example deliberately contrived to exhibit the relationships involved. Suppose we have a pair of dice that are rolled together. If the first die comes to rest with side 6 on the top, that is an event of the type A; if the

---

[3] Because only two effects, A and B, appear in formulas (1)–(4), I mention only two individuals in this example. The definition of the conjunctive fork can obviously be generalized to handle a larger number of cases.

[4] If other potential common causes exist, we can form a partition, $C_1, C_2, \ldots, C_n, \bar{C}$, and the corresponding relations will obtain. Equation (1) would be replaced by

$$P(A.B|C_i) = P(A|C_i) \times P(B|C_i)$$

and equations (3) and (4) would be replaced by

$$P(A|C_i) > P(A|\bar{C})$$
$$P(B|C_i) > P(B|\bar{C}).$$

second die comes to rest with side 6 uppermost, that is an event of type
B. These dice are like standard dice except for the fact that each one has
a tiny magnet embedded in it. In addition, the table on which they are
thrown has a powerful electromagnet beneath its surface. This magnet can
be turned on or off with a concealed switch. If the dice are rolled when
the electromagnet is on, it is considered an instance of the common cause
C; if the magnet is off when the dice are tossed, the event is designated
as $\bar{C}$. Let us further assume that when the electromagnet is turned off,
these dice behave exactly as standard dice. The probability of getting 6
with either die is 1/6, and the probability of getting double 6 is 1/36.[5] If
the electromagnet is turned on, let us assume, the chance of getting 6 with
either die is 1/2, and the probability of double 6 is 1/4. It is easily seen
that conditions (1)–(4) are fulfilled. Let us make a further stipulation,
which will simplify the arithmetic, but which has no other bearing upon
the essential features of the example—namely, that half of the tosses of
this pair of dice are made with the electromagnet turned on, and half are
made with it turned off. We might imagine some sort of random device
that controls the switch, and that realizes this equiprobability condition.
We can readily see that the overall probability of 6 on each die, regardless
of whether the electromagnet is on or off, is 1/3. In addition, the overall
probability of double 6 is the arithmetical average of 1/4 and 1/36, which
equals 5/36. If the occurrence of 6 on one die were independent of 6
occurring on the other, the overall probability of double 6 would be 1/3
$\times$ 1/3 = 1/9 $\neq$ 5/36. Thus the example satisfies relation (5), as of course
it must, in addition to relations (1)–(4).

It may initially seem counterintuitive to say that the results on the two
dice are statistically independent if the electromagnet is off, and they are
statistically independent if it is on, but that overall they are not independent.
But they are, indeed, nonindependent, and this nonindependence arises
from a clustering of 6s, which is due simply to the fact that in a subset of
the class of all tosses the probability of 6 is enhanced for both dice. The
dependency arises, not because of any physical interaction between the
dice, but because of special background conditions that obtain on certain
of the tosses but not on others. The same consideration applies to the
earlier, less contrived, cases. When the two students each copy from a
paper in a fraternity file, there is no direct physical interaction between
the process by which one of the papers is produced and that by which the
other is produced—in fact, if either student had been aware that the other
was using that source, the unhappy coincidence might have been avoided.
Likewise, as explicitly mentioned in the mushroom poisoning case—where,

[5] I am assuming that the magnet in one die does not affect the behavior of the other die.

to make the example fit formulas (1)–(4), we confine attention to just two of the performers—the illness of one of them had no effect upon the illness of the other. The coincidence resulted from the fact that a common set of background conditions obtained, namely, a common food supply from which both ate. Similarly, in the twin quasar example, the two images are formed by two separate radiation processes that come from a common source, but do not directly interact with each other anywhere along the line.

Reichenbach claimed—correctly, I believe—that conjunctive forks possess an important asymmetry. Just as we can have two effects that arise out of a given common cause, so also may we find a common effect resulting from two distinct causes. For example, by getting results on a roll of two dice that add up to 7, one may win a prize. Reichenbach distinguished three situations: (1) a common cause C giving rise to two separate effects, A and B, without any common effect arising from A and B conjointly; (2) two events A and B that, in the absence of a common cause C, jointly produce a common effect E; and (3) a combination of (1) and (2) in which the events A and B have both a common cause C and a

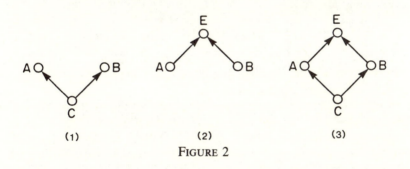

FIGURE 2

common effect E. He characterized situations (1) and (2) as *open forks*, while (3) is closed on both ends. Reichenbach's asymmetry thesis was that situations of type (2) never represent conjunctive forks; conjunctive forks that are open are always open to the future and never to the past. Since the statistical relations found in conjunctive forks are said to explain otherwise improbable coincidences, it follows that such coincidences are explained only in terms of common causes, never common effects. In the case of a prize being awarded for the result 7 on a toss of two dice, we do not explain the occurrence of 7 in terms of the awarding of the prize. This is not a mere philosophical prejudice against teleological explanations. Assuming a fair game, we believe, in fact, that the probability of getting a 7 is the same regardless of whether a prize is involved. In situations of type (2), in which there is no common cause to produce a statistical dependency, A and B occur independently; the common effect E, unlike a common cause, does not create a correlation between A and B. This is a straightforward factual assertion, which can, in cases like the tossing of dice, be tested empirically. If, contrary to expectation, we should find that the result 7 does not occur with a probability of 1/6 in cases where a prize is at stake, we could be confident that the (positive or negative) correlation between the outcomes on the two dice was a result of some prior tampering—recall the magnetic dice, where the electromagnet had to be turned on before the dice came to rest to affect the probability of the result—rather than 'events conspiring' to reward one player or to prevent another from receiving a benefit. A world in which teleological causation operates is not logically impossible, but our world does not seem, as a matter of fact, to be of such a kind.

In order to appreciate fully the import of Reichenbach's asymmetry thesis, let us look at an initially plausible putative counterexample provided by Frank Jackson.[6] It will be instructive to make an explicit comparison between his example and a bona fide instance of a common cause. Let us begin with the common cause. Suppose that two siblings contract mumps at the same time, and assume that neither caught the disease from the other. The coincidence is explained by the fact that they attended a birthday party and, by virtue of being in the same locale, both were exposed to another child who had the disease. This would constitute a typical example of a conjunctive fork.

Now, with that kind of example in mind, consider a case that involves Hansen's disease (leprosy). One of the traditional ways of dealing with this illness was by segregating its victims in colonies. Suppose that Adams has Hansen's disease (A) and Baker also has it (B). Previous to contracting the disease, Adams and Baker had never lived in proximity to one another, and there is no victim of the disease with whom both had been in contact. We may therefore assume that there is no common cause. Subsequently, however, Adams and Baker are transported to a colony, where both are treated with chaulmoogra oil (the traditional treatment). The fact that both Adams and Baker are in the colony and exposed to chaulmoogra oil is a common effect of the fact that each of them has Hansen's disease. This situation, according to Jackson, constitutes a conjunctive fork A, E, B, where we have a common effect E, but no common cause. We must see whether it does, in fact, qualify. It is easy to see that relations (3) and (4) are satisfied, for

[6] This example was offered at a meeting of the Victoria Section of the Australasian Association for History and Philosophy of Science at the University of Melbourne in 1978.

$$P(A|E) > P(A|\bar{E})$$

and

$$P(B|E) > P(B|\bar{E}),$$

that is, the probability of Adams having Hansen's disease is greater if he and Baker are members of the colony than it would be if they were not. If not both Adams and Baker are members of the colony, it might be that Adams is a member and Baker is not (in which case the probability that Adams has the disease is high), but it might also be that Baker is a member and Adams is not, or that neither of them is (in which cases the probability that Adams has the disease would be very low). The same reasoning holds for Baker, mutatis mutandis.

The crucial question concerns relations (1) and (2). Substituting "E" for "C" in those two equations, let us recall that they say, respectively, that A and B are statistically independent of one another, given that condition E holds, and they are statistically independent when E does not hold. Now, if A and B are independent of one another, it follows immediately that their negations $\bar{A}$ and $\bar{B}$ are independent; thus relation (1) implies

$$P(\bar{A}.\bar{B}|E) = P(\bar{A}|E) \times P(\bar{B}|E).$$

This tells us, as we saw previously, that E screens off $\bar{A}$ from $\bar{B}$, that is,

$$P(\bar{B}|E) = P(\bar{B}|\bar{A}.E).$$

Let us therefore ask whether the fact that both Adams and Baker are members of the colony (and both are exposed to chaulmoogra oil) would make the fact that Adams did not have Hansen's disease statistically irrelevant to Baker's failure to have that disease. The answer is clearly negative. Among the members of the colony, a small percentage—doctors, nurses, missionaries—do not have Hansen's disease, and those involved in actual treatment of the victims are exposed to chaulmoogra oil. Suppose, for example, that Adams and Baker both belong to the colony and are exposed to chaulmoogra oil, but that Baker does not have leprosy. To make the situation concrete, suppose that there are one hundred members of the colony who are exposed to chaulmoogra oil, and among them are only two medical personnel who do not have Hansen's disease. If Baker has Hansen's disease, the probability that Adams does not have it is about 0.02, while if Baker does not have it, the probability for Adams is about 0.01—a difference of a factor of two. As stipulated in this example, the fact that Adams has the disease has no direct causal relevance to the fact that Baker also has it, and conversely, but given the circumstances specified, they are statistically relevant to one another.

Although we know that A, B, and E do not form a conjunctive fork, let us look also at relation (2). For this purpose, we shall ask whether $\bar{E}$ screens off A and B from each other—that is, whether

$$P(B|\bar{E}) = P(B|A.\bar{E}).$$

Let us suppose, for the sake of this argument, that there is only one colony for victims of Hansen's disease, and that almost every afflicted person belongs to it—indeed, let us assume for the moment that only two people in the world have Hansen's disease who have not yet joined the colony. Thus, if a pair of people are chosen at random, and if not both belong to the colony, it is quite unlikely that even one of them has Hansen's disease. However, among the many pairs not both of whom are in the colony, but at least one of whom has Hansen's disease, there is only one consisting of two individuals who both have this disease. Thus, under our extreme temporary assumption about the colony, it is clear that $\bar{E}$ does not screen off A from B. Given that not both Adams and Baker are members of the colony receiving the traditional treatment, it is quite unlikely that Baker has Hansen's disease, but given in addition that Adams has the disease, it becomes even more unlikely that Baker has it. The assumption that there are only two victims of Hansen's disease in the world who are not members of the colony is, of course, altogether unrealistic; however, given the relative rarity of the disease among the population at large, the failure of screening off, though not as dramatic, must still obtain. We see from the failure of relations (1) and (2) that Jackson's example does not constitute a conjunctive fork. In spite of the great superficial similarity between the mumps and leprosy situations, there is a deep difference between them. This comparison exemplifies Reichenbach's asymmetry thesis.

Although Reichenbach held only that there are no conjunctive forks that are open toward the past, I believe that an even stronger claim is warranted—though I shall merely illustrate it but not try to argue it here. I am inclined to believe that conjunctive forks, whether open or closed by a fourth event, always point in the same temporal direction. Reichenbach allowed that in situations of type (3), the two events A and B along with their common effect E could form a conjunctive fork. Here, of course, there must also be a common cause C, and it is C rather than E that explains the coincidental occurrence of A and B. I doubt that, even in these circumstances, A, B, and E can form a conjunctive fork.

Suppose—to return to the mushroom poisoning example mentioned previously—that among the afflicted troupers are the leading lady and leading man, and that the performance scheduled for that evening is canceled. I shall assume, as before, that the two illnesses along with the common cause form a conjunctive fork. The analysis that shows that the fork consisting of the two illnesses and the cancellation of the performance is not conjunctive is similar to the analysis of the leprosy example. Let us focus upon equation (2), and consider it from the standpoint of screening off. We must compare the probability that the leading lady is ill, given that the performance is not canceled, with the probability that the leading lady is ill, given that the leading man is ill and the performance is not canceled. It is implausible to suppose that the two are equal, for it seems much more likely that the show could go on if only one of the leading characters is ill than it would be if both were ill. It should be recalled that we are discussing a traveling theatrical company, so we should not presume that a large complement of stand-ins is available. Therefore, the performance of the play does not screen off the one illness from the other, and relation (2) in the definition of the conjunctive fork does not hold. One could use a similar argument to show, in this case, that relation (1) is also violated.

I do not have any compelling argument to prove that in the case of the double fork (Fig. 2-(3)), A, C, B and A, E, B cannot both be conjunctive. Such combinations are logically possible, for one can find sets of probability values that satisfy the eight relations—four for each fork.[7] Nevertheless, I have not been able to find or construct a physically plausible example, and Reichenbach never furnished one. It would be valuable to know whether double conjunctive forks of this sort would violate some basic physical principle. I do not know the answer to this question.

Reichenbach's principle of the common cause asserts the existence of common causes in cases of improbable coincidences, but it does not assert that *any* event C that fulfills relations (1)–(4) qualifies ipso facto as a common cause of A and B.[8] The following example, due to Ellis Crasnow,

---

[7] As a matter of fact, Crasnow's example, described in the next paragraph, illustrates this point. In that example, we have two events A and B that form a conjunctive fork with an earlier event C, and also with another earlier event C'. Hence the four events C', A, B, and C form a double conjunctive fork with vertices at C' and C. In this example, C' qualifies as a bona fide common cause, but C is a spurious common cause. Since C precedes A and B, it cannot be a common effect of those two events; nevertheless, the four events fulfill all of the mathematical relations that define conjunctive forks, which shows that double conjunctive forks are logically possible. If it is difficult—or impossible—to find double conjunctive forks that include both a bona fide common cause and a bona fide common effect, the problem is one of physical, rather than mathematical, constraints. The fact that C occurs before A and B rather than after is a physical fact upon which the probability relations in formulas (1)–(4) have no bearing.

[8] In (Salmon, 1980), I took Crasnow's example as a counterexample to Reichenbach's theory, but, as Paul Humphreys kindly pointed out in a private communication, this was an error.

illustrates this point. Consider a man who usually arrives at his office about 9:00 A.M., makes a cup of coffee, and settles down to read the morning paper. On some occasions, however, he arrives promptly at 8:00 A.M., and on these very same mornings his secretary has arrived somewhat earlier and prepared a fresh pot of coffee. Moreover, on just these mornings, he is met at his office by one of his associates who normally works at a different location. Now, if we consider the fact that the coffee is already made when he arrives (A) and the fact that his associate shows up on that morning (B) as the coincidence to be explained, then it might be noted that on such mornings he always catches the 7:00 A.M. bus (C), while on other mornings he usually takes the 8:00 A.M. bus ($\bar{C}$). In this example, it is plausible enough to suppose that A, B, and C form a conjunctive fork satisfying (1)–(4), but obviously C cannot be considered a cause either of A or of B. The actual common cause is an entirely different event C', namely, a telephone appointment made the day before by his secretary. C' is, in fact, the common cause of A, B, and C.

This example leaves us with an important question. Given an event C that, along with A and B, forms a conjunctive fork, how can we tell whether C is a bona fide common cause? The answer, I think, is quite straightforward. C must be connected to A and to B by suitable causal processes of the sort discussed in the preceding chapter. These causal processes constitute the mechanisms by which causal influence is transmitted from the cause to each of the effects.

INTERACTIVE FORKS

There is another, basically different, sort of common cause situation that cannot appropriately be characterized in terms of conjunctive forks. Consider a simple example. Two pool balls, the cue ball and the 8-ball, lie upon a pool table. A relative novice attempts a shot that is intended to put the 8-ball into one of the far corner pockets, but given the positions of the balls, if the 8-ball falls into one corner pocket, the cue ball is almost certain to go into the other far corner pocket, resulting in a 'scratch.' Let A stand for the 8-ball dropping into the one corner pocket, let B stand for the cue ball dropping into the other corner pocket, and let C stand for the collision between the cue ball and the 8-ball that occurs when the player executes the shot. We may reasonably assume that the probability of the 8-ball going into the pocket is 1/2 if the player tries the shot, and that the probability of the cue ball going into the pocket is also about 1/2. It is immediately evident that A, B, and C do not constitute a conjunctive fork, for C does not screen off A and B from one another. Given that the shot is attempted, the probability that the cue ball will fall into the pocket (approximately 1/2) is not equal to the probability that the cue ball will go into the pocket, given that the shot has been attempted and that the 8-ball has dropped into the other far corner pocket (approximately 1).

In discussing the conjunctive fork, I took some pains to point out that forks of that sort occur in situations in which separate and distinct processes, which do not directly interact, arise out of special background conditions. In the example of the pool balls, however, there is a direct interaction—a collision—between the two causal processes consisting of portions of the histories of the two balls. For this reason, I have suggested that forks that are exemplified by such cases be called *interactive forks*.[9] Since the common cause C does not statistically screen off the two effects A and B from one another, interactive forks violate condition (1) in the definition of conjunctive forks.

The best way to look at interactive forks, I believe, is in terms of spatiotemporal intersections of processes. In some cases, two processes may intersect without producing any lasting modification in either. This will happen, for example, when both processes are pseudo-processes. If the paths of two airplanes, flying in different directions at different altitudes on a clear day, cross one another, the shadows on the ground may coincide momentarily. But as soon as the shadows have passed the intersection, both move on as if no such intersection had ever occurred. In the case of the two pool balls, however, the intersection of their paths results in a change in the motion of each that would not have occurred if they had not collided. Energy and momentum are transferred from one to the other; their respective states of motion are altered. Such modifications occur, I shall maintain, only when (at least) two causal processes intersect. If either or both of the intersecting processes are pseudo-processes, no such mutual modification occurs. However, it is entirely possible for two causal processes to intersect without any subsequent modification in either. Barring the extremely improbable occurrence of a particle-particle type collision between two photons, light rays normally pass right through one another without any lasting effect upon either one of them. The fact that two intersecting processes are both causal is a necessary but not sufficient condition of the production of lasting changes in them.

When two causal processes intersect and suffer lasting modifications after the intersection, there is some correlation between the changes that occur in them. In many cases—and perhaps all—energy and/or momentum transfer occurs, and the correlations between the modifications are direct

---

[9] I am deeply indebted to Philip von Bretzel (1977, note 13) for the valuable suggestion that causal interactions might be explicated in terms of causal forks. For further elaboration of the relations between the two kinds of forks, see (Salmon, 1978).

consequences of the respective conservation laws.[10] This is illustrated by the Compton scattering of an energetic photon off of an electron that can be considered, for practical purposes, initially at rest. The difference in energy between the incoming photon $h\nu$ and the scattered photon $h\nu'$ is equal to the kinetic energy of the recoiling electron. Similarly, the momentum change in the photon is exactly compensated by the momentum change in the electron.[11]

When two processes intersect, and they undergo correlated modifications that persist after the intersection, I shall say that the intersection constitutes a *causal interaction*. This is the basic idea behind what I want to take as a fundamental causal concept. Let C stand for the event consisting of the intersection of two processes. Let A stand for a modification in one and B for a modification in the other. Then, in many cases, we find a relation analogous to equation (1) in the definition of the conjunctive fork, except that the equality is replaced by an inequality:

$$P(A.B|C) > P(A|C) \times P(B|C). \tag{9}$$

Moreover, given a causal interaction of the foregoing sort, I shall say that the change in each process is *produced* by the interaction with the other process.

I have now characterized, at least partially, the two fundamental causal concepts mentioned at the beginning of the preceding chapter. Causal processes are the means by which causal influence is *propagated*, and changes in processes are *produced* by causal interactions. We are now in a position to see the close relationship between these basic notions. The distinction between causal processes and pseudo-processes was formulated in terms of the criterion of mark transmission. A mark is a modification in a process, and if that modification persists, the mark is transmitted. Modifications in processes occur when they intersect with other processes; if the modifications persist beyond the point of intersection, then the intersection constitutes a causal interaction and the interaction has produced marks that are transmitted. For example, a pulse of white light is a process, and a piece of red glass is another process. If these two processes intersect—that is, if the light pulse goes through the red glass—then the light pulse becomes and remains red, while the filter undergoes an increase in energy as a result of absorbing some of the light that impinges upon it. Although the newly acquired energy may soon be dissipated into the surrounding

---

[10] For an important discussion of the role of energy and momentum transfer in causality, see (Fair, 1979).

[11] As explained in (Salmon, 1978), the example of Compton scattering has the advantage of being irreducibly statistical, and thus, not analyzable, even in principle, as a perfect fork (discussed in a subsequent section of this chapter).

environment, the glass retains some of the added energy for some time beyond the actual moment of interaction.

We may, therefore, turn the presentation around in the following way. We live in a world which is full of processes (causal or pseudo), and these processes undergo frequent intersections with one another. Some of these intersections constitute causal interactions; others do not. Let us attempt to formulate a principle CI (for causal interaction) that will set forth the condition explicitly:

CI: *Let P_1 and P_2 be two processes that intersect with one another at the space-time point S, which belongs to the histories of both. Let Q be a characteristic that process P_1 would exhibit throughout an interval (which includes subintervals on both sides of S in the history of P_1) if the intersection with P_2 did not occur; let R be a characteristic that process P_2 would exhibit throughout an interval (which includes subintervals on both sides of S in the history of P_2) if the intersection with P_1 did not occur. Then, the intersection of P_1 and P_2 at S constitutes a causal interaction if:*

(1) P_1 exhibits the characteristic Q before S, but it exhibits a modified characteristic Q' throughout an interval immediately following S; and

(2) P_2 exhibits the characteristic R before S, but it exhibits a modified characteristic R' throughout an interval immediately following S.

The modifications that Q and R undergo will normally—perhaps invariably—be correlated to one another in accordance with some conservation law, but it seems unnecessary to include this as a requirement in the definition.

This principle, like the principle MT (mark transmission), is formulated in counterfactual terms, for reasons similar to those that induced us to employ a counterfactual formulation in MT. Consider the following example. Two spots of light, one red and the other green, are projected on a white screen. The red spot moves diagonally across the screen from the lower left-hand corner to the upper right-hand corner, and the green spot moves diagonally from the lower right-hand corner to the upper left-hand corner. Let these spots be projected in such a way that they meet and merge momentarily at the center of the screen; at that moment, a yellow spot appears (the result of combining red and green light—mixing colored light is altogether different from mixing colored paints), but each resumes its former color as soon as it leaves the region of intersection. Since no modification of color persists beyond the intersection, we are not tempted to suppose, on the basis of this observation, that a causal interaction has occurred.

Now let us modify the setup. Again, we have the two spots of light

172 | Causal Forks and Common Causes

projected upon the screen, but in the new situation they travel in different paths. The red spot moves diagonally from the lower left-hand corner to the center of the screen, and then it travels from the center to the upper left-hand corner. The green spot moves from the lower right-hand corner to the center, and then to the upper right-hand corner. Assuming, as before, that the two spots of light meet at the center of the screen, we could describe what we see in either of two ways. First, we could say that the red spot collides with the green spot in the center of the screen, and the directions of motion of the two spots are different after the collision. Second, we could say that the spot that travels from the lower left to the upper right changes from red to green as it goes through the intersection in the middle of the screen, while the spot that travels from lower right to upper left changes from green to red as it goes through the intersection. It seems that each spot changes color in the intersection and the change persists beyond the locus of the intersection. Under either of these descriptions, it may appear that we have observed a causal interaction, but such an appearance is illusory. Two pseudo-processes intersected, but no causal interaction occurred.

The counterfactual formulation of principle CI is designed to deal with examples of this sort. Under the conditions specified in the physical setup, the red spot moves from the lower left corner to the center and then to the upper left corner of the screen regardless of whether the green spot is present or not. It would be false to say that the red spot would have traveled across the screen from lower left to upper right if it had not met the green spot. Parallel remarks apply, mutatis mutandis, to the behavior of the green spot. Similarly, it would be false to say that the color of the spot that traveled from lower left to upper right would not have changed color if it had not encountered the spot traveling from lower right to upper left. CI is not vulnerable to putative counterexamples of this sort.

Examples of another sort, which were presented independently by Patrick Maher and Richard Otte, are more difficult to handle. Suppose (to use Otte's version) that two billiard balls roll across a table with a transparent surface and that they collide with one another in the center, with the result that their directions of motion are changed. This is, of course, a bona fide causal interaction, and it qualifies as such under CI. We are entitled to say that the direction of motion of the one ball would not have changed in the middle of the table if the collision with the second had not occurred. It is easy to see how that counterfactual could be tested in a controlled experiment. Assume, further, that because of a bright light above the table the billiard balls cast shadows on the floor. When the balls collide the shadows meet, and their directions of motion are different after the intersection. In this case, we would appear to be entitled to say that the direction of motion of the one shadow would not have changed if it had not encountered the other shadow. It looks as if the intersection of the two shadows qualifies as a causal interaction according to CI.

In order to handle examples of this kind, we must consider carefully how our counterfactuals are to be interpreted. This issue arose in connection with the principle of mark transmission (MT) in chapter 5, and we made appeal in that context to the testing of counterfactual assertions by means of controlled experiments. We must, I think, approach the question in the same way in this context. We must therefore ask what kind of controlled experiment would be appropriate to test the assertion about the shadows of the colliding billiard balls.

If we just sit around watching the shadows, we will notice some cases in which a shadow, encountering no other shadows, moves straight along the floor. We will notice other cases in which two shadows intersect, and their directions of motions are different after the intersection than they were before. So we can see a correlation between alterations of direction of motion and intersections with other shadows. This procedure, however, consists merely of the collection of available data; it hardly constitutes experimentation. So we must give further thought to the design of some experiments to test our counterfactuals.

Let us formulate the assertion that is to be tested as follows: If shadow #1 had not met shadow #2, then shadow #1 would have continued to move in a straight line instead of changing directions. In order to perform a controlled experiment, we need a situation in which the phenomena under investigation occur repeatedly. Let us mark off a region of the floor that is to be designated as the experimental region. Assume that we have many cases in which two shadows enter that region moving in such a way that they will meet within it, and that their directions of motion after the intersection are different from their prior directions of motion. How this is to be accomplished is no mystery. We could study the behavior of the shadows that occur as many games of billiards are played, or we could simply arrange for an experimenter to roll balls across the table in such a way that they collide with one another in the desired place. Consider one hundred such occurrences. Using some random device, we select fifty of them to be members of the experimental group and fifty to be members of the control group. That selection is not communicated to the experimenter who is manipulating the balls on the table top. When an event in the control group occurs, we simply do nothing and observe the outcome. When an event in the experimental group occurs, we choose one of the entering shadows—call it shadow #2—and shine a light along the path it would have taken if the light had not been directed toward it. This extra illumination obliterates shadow #2; nevertheless, shadow #1 changes its direction in the experimental cases just as it does in the control cases. We have thereby established the falsity of the counterfactual that was to be tested. It is not true that the direction of travel of shadow #1 would not have changed if it had not encountered shadow #2. We have fifty instances in which shadow #1 changed its direction in the absence of shadow #2. The intersection of the two shadows thus fails, according to CI, to qualify as a causal interaction.

As formulated, CI states only a sufficient condition for a causal interaction, since there might be other characteristics, F and G, that suffer the requisite mutual modification even if Q and R do not. In order to transform CI into a condition that is necessary as well as sufficient, we need simply to say that a causal interaction occurs if and only if there exist characteristics Q and R that fulfill the conditions stated previously. It should be noted that the statistical relation (9), which may be a true statement about a given causal interaction, does not enter into the definition of causal interactions.[12]

If two processes intersect in a manner that qualifies as a causal interaction, we may conclude that both processes are causal, for each has been marked (i.e., modified) in the intersection with the other and each process transmits the mark beyond the point of intersection. Thus each fulfills the criterion MT; each process shows itself capable of transmitting marks since each one has transmitted a mark generated in the intersection. Indeed, the operation of marking a process is accomplished by means of a causal interaction with another process. Although we may often take an active role in producing a mark in order to ascertain whether a process is causal (or for some other purpose), it should be obvious that human agency plays no essential part in the characterization of causal processes or causal interactions. We have every reason to believe that the world abounded in causal processes and causal interactions long before there were any human agents to perform experiments.

RELATIONS BETWEEN CONJUNCTIVE AND INTERACTIVE FORKS

Suppose that we have a shooting gallery with a number of targets. The famous sharpshooter Annie Oakley comes to this gallery, but it presents no challenge to her, for she can invariably hit the bull's-eye of any target at which she aims. So, to make the situation interesting, a hardened steel knife-edge is installed in such a position that a direct hit on the knife-edge will sever the bullet in a way that makes one fragment hit the bull's-eye

---

[12] In (Salmon, 1978), I suggested that interactive forks could be defined statistically, in analogy with conjunctive forks, but I now think that the statistical characterization is inadvisable.

of target A while the other fragment hits the bull's-eye of target B. If we let A stand for a fragment striking the bull's-eye of target A, B for a fragment striking the bull's-eye of target B, and C for the severing of the bullet by the knife-edge, we have an interactive fork quite analogous to the example of the pool balls. Indeed, we may use the same probability values, setting $P(A|C) = P(B|C) = 1/2$, while $P(A|C.B) = P(B|C.A) \simeq 1$. Statistical screening off obviously fails.

We might, however, consider another event C^*. To make the situation concrete, imagine that we have installed between the knife-edge and the targets a steel plate with two holes in it. If the shot at the knife-edge is good, then the two fragments of the bullet will go through the two holes, and each fragment will strike its respective bull's-eye with probability virtually equal to 1. Let C^* be the event of the two fragments going through their respective holes. Then, we may say, A, B, and C^* will form a conjunctive fork. That happens because C^* refers to a situation that is subsequent to the physical interaction between the parts of the bullet. By the time we get to C^*, the bullet has been cut into two separate pieces, and each is going its way independently of the other. Even if we should decide to vaporize one of the fragments with a powerful laser, that would have no effect upon the probability of the other fragment finding its target. This example makes quite vivid, I believe, the distinction between the interactive fork, which characterizes direct physical interactions, and the conjunctive fork, which characterizes independent processes arising under special background conditions.[13]

There is a further important point of contrast between conjunctive and interactive forks. Conjunctive forks possess a kind of temporal asymmetry, which was described previously. Interactive forks do not exhibit the same sort of temporal asymmetry. This is easily seen by considering a simple collision between two billiard balls. A collision of this type can occur in reverse; if a collision C precedes states of motion A and B in the two balls, then a collision C can occur in which states of motion just like A and B, except that the direction of motion is reversed, precede the collision. Causal

[13] In an article entitled "When are Probabilistic Explanations Possible?" Suppes and Zanotti begin with the assertion: "The primary criterion of adequacy of a probabilistic causal analysis is that the causal variable should render the simultaneous phenomenological data conditionally independent. The intuition back of this idea is that the common cause of the phenomena should factor out the observed correlations. So we label the principle the *common cause criterion*" (1981, p. 191, italics in original). This statement amounts to the claim that all common cause explanations involve conjunctive forks; they seem to overlook the possibility that interactive forks may be involved. One could, of course, attempt to defend this principle on the ground that for any interactive cause C, it is possible to find a conjunctive cause C^*. While this argument may be acceptable for macroscopic phenomena, it does not seem plausible for such microscopic phenomena as Compton scattering.

interactions and causal processes do not, in and of themselves, provide a basis for temporal asymmetry.

Our ordinary causal language is infused with temporal asymmetry, but we should be careful in applying it to basic causal concepts. If, for example, we say that two processes are modified as a result of their interaction, the words suggest that we have already determined which are the states of the processes prior to the interaction, and which are the subsequent states. To avoid begging temporal questions, we should say that two processes intersect, and each of the processes had different characteristics on the two sides of the intersection. We do not try to say which part of the process came earlier and which later.[14] The same is true when we speak of marking. To erase a mark is the exact temporal reverse of imposing a mark; to speak of imposing or erasing is to presuppose a temporal direction. In many cases, of course, we know on other grounds that certain kinds of interactions are irreversible. Light filters absorb some frequencies, so that they transform white light into red. Filters do not furnish missing frequencies to turn red light into white. But until we have gone into the details of the physics of irreversible processes, it is best to think of causal interactions in temporally symmetric terms, and to take the causal connections furnished by causal processes as symmetric connections. Causal processes and causal interactions do not furnish temporal asymmetry; conjunctive forks fulfill that function.

It has been mentioned in this chapter that the cause C in an interactive fork does not *statistically* screen off the effect A from the effect B. There is, however, a kind of *causal* screening off that is a feature of macroscopic interactive forks.[15] In order for ABC to form an interactive fork, there must be causal processes connecting C to A and C to B. Suppose we mark the process that connects C to A at some point between C and A. If we do not specify the relations of temporal priority, then the mark may be transmitted to A or it may be transmitted to C. But whichever way it goes, the mark will not be transmitted to the process connecting C to B. Similarly, a mark imposed upon the process connecting C to B will not be transmitted to the

[14] The principle CI, as formulated previously, involves temporal commitments of just this sort. However, these can be purged easily by saying that P_1 and P_2 exhibit Q and R, respectively, on one side of the intersection at S, and they exhibit Q' and R', respectively, on the other side of S. With this reformulation, CI becomes temporally symmetric. When one is dealing with questions of temporal anisotropy or 'direction,' this symmetric formulation should be adopted. Problems regarding the structure of time are not of primary concern in this book; nevertheless, I am trying to develop causal concepts that will fit harmoniously with a causal theory of time.

[15] As Bas van Fraassen kindly pointed out at a meeting of the Philosophy of Science Association, the restriction to macroscopic cases is required by the kinds of quantum phenomena that give rise to the Einstein-Podolsky-Rosen problem.

process connecting C to A. This means that no causal influence can be transmitted from A to B or from B to A via C. C constitutes an effective causal barrier between A and B even if A and B exhibit the sort of statistical correlation formulated in (9). The same kind of causal screening occurs in the conjunctive fork, of course, but in forks of this type C statistically screens off A from B as well.

It may be worth noting that marks can be transmitted through interactions, so that if A, C, and B have a linear causal order, a mark made in the process connecting A with C may be transmitted through C to B. For example, if A stands for the impulsion of a cue ball by a cue stick, C for the collision of the cue ball with the 8-ball, and B for the cue ball dropping into the pocket, then it may happen that the tip of the cue stick leaves a blue chalk mark on the cue ball, and that this mark remains on the cue ball until it drops into the pocket. More elaborate examples are commonplace. If a disk jockey at a radio station plays a record that has a scratch on its surface, the mark will persist through all of the many physical processes that take place between the contact of the stylus with the scratch and the click that is perceived by someone who is listening to that particular station.

Perfect Forks

In dealing with conjunctive and interactive forks, it is advisable to restrict our attention to the cases in which $P(A|C)$ and $P(B|C)$ do not assume either of the extreme values of zero or one. The main reason is that the relation

$$P(A.B|C) = P(A|C) \times P(B|C) = 1 \tag{10}$$

may represent a limiting case of either a conjunctive or an interactive fork, even though (10) is a special case of equation (1) and it violates relation (9).

Consider the Annie Oakley example once more. Suppose that she returns to the special shooting gallery time after time. Given that practice makes perfect (at least in her case), she improves her skill until she can invariably hit the knife-edge in the manner that results in the two fragments finding their respective bull's-eyes. Up until the moment that she has perfected her technique, the results of her trials exemplified interactive forks. It would be absurd to claim that when she achieves perfection, the splitting of the bullet no longer constitutes a causal interaction, but must now be regarded as a conjunctive fork. The essence of the interactive fork is to achieve a high correlation between two results; if the correlation is perfect, we can ask no more. It is, one might say, an arithmetical accident that when perfection occurs, equation (1) is fulfilled while the inequality (9) must be violated. If probability values were normalized to some value other than 1, that result would not obtain. It therefore seems best to treat this special case as a third type of fork—the *perfect fork*.

Conjunctive forks also yield perfect forks in the limit. Consider the example of illness due to consumption of poisonous mushrooms. If we assume—what is by no means always the case—that anyone who consumes a significant amount of the mushroom in question is certain to become violently ill, then we have another instance of a perfect fork. Even when these limiting values obtain, however, there is still no direct interaction between the processes leading respectively to the two cases of severe gastrointestinal distress.

The main point to be made concerning perfect forks is that when the probabilities take on the limiting values, it is impossible to tell from the statistical relationships alone whether the fork should be considered interactive or conjunctive. The fact that relations (1)–(4), which are used in the characterization of conjunctive forks, are satisfied does not constitute a sufficient basis for making a judgment about the temporal orientation of the fork. Only if we can establish, on separate grounds, that the perfect fork is a limiting case of a conjunctive (rather than an interactive) fork, can we conclude that the event at the vertex is a common cause rather than a common effect.[16] Perfect forks need to be distinguished from the other two types mainly to guard against this possible source of confusion.

The Causal Structure of the World

In everyday life, when we talk about cause-effect relations, we think typically (though not necessarily invariably) of situations in which one event (which we call the cause) is linked to another event (which we call the effect) by means of a causal process. Each of the two events in this relation is an interaction between two (or more) intersecting processes. We say, for example, that the window was broken by boys playing baseball. In this situation, there is a collision of a bat with a ball (an interactive fork), the motion of the ball through space (a causal process), and a collision of the ball with the window (an interactive fork). We say, for another example, that turning a switch makes the light go on. In this case, an interaction between a switching mechanism and an electrical circuit leads to a process consisting of a motion of electric charges in some wires, which in turn leads to emission of light from a filament. Homicide by shooting

---

[16] It must be an open rather than a closed fork. In suggesting previously that all conjunctive forks have the same temporal orientation, it was to be understood that we were talking about bona fide conjunctive forks, not limiting cases that qualify as perfect forks.

provides still another example. An interaction between a gun and a cartridge propels a bullet (a causal process) from the gun to the victim, where the bullet then interacts with the body of the victim.

The foregoing characterization of causal processes and various kinds of causal forks provides, I believe, a basis for understanding three fundamental aspects of causality:

1. *Causal processes* are the means by which structure and order are *propagated* or transmitted from one space-time region of the universe to other times and places.

2. *Causal interactions*, as explicated in terms of interactive forks, constitute the means by which *modifications in structure* (which are propagated by causal processes) are *produced*.

3. Conjunctive *common causes*—as characterized in terms of conjunctive forks—play a vital role in the *production* of structure and order. In the conjunctive fork, it will be recalled, two or more processes, which are physically independent of one another and which do not interact directly with each other, arise out of some special set of background conditions. The fact that such special background conditions exist is the source of a correlation among the various effects that would be utterly improbable in the absence of the common causal background.

There is a striking difference between conjunctive common causes on the one hand and causal processes and interactions on the other. Causal processes and causal interactions seem to be governed by basic laws of nature in ways that do not apply to conjunctive forks. Consider two paradigms of causal processes, namely, an electromagnetic wave propagating through a vacuum and a material particle moving without any net external forces acting upon it. Barring any causal interactions in both cases, the electromagnetic wave is governed by Maxwell's equations and the material particle is governed by Newton's first law of motion (or its counterpart in relativity theory). Causal interactions are typified by various sorts of collisions. The correlations between the changes that occur in the processes involved are governed—in most, if not all, cases—by fundamental physical conservation laws.

Conjunctive common causes are not nearly as closely tied to the laws of nature. It should hardly require mention that to the extent that conjunctive forks involve causal processes and causal interactions, the laws of nature apply as sketched in the preceding paragraph. However, in contrast to causal processes and causal interactions, conjunctive forks depend crucially upon de facto background conditions. Recall some of the examples mentioned previously. In the plagiarism example, it is a nonlawful fact that two members of the same class happen to have access to the same file of term papers. In the mushroom poisoning example, it is a nonlawful fact that the players sup together out of a common pot. In the twin quasar example, it is a de facto condition that the quasar and the elliptical galaxy are situated in such a way that light coming to us from two different directions arises from a source that radiates quite uniformly from extended portions of its surface.

There is a close parallel between what has just been said about conjunctive forks and what philosophers like Reichenbach (1956, chap. 3) and Grünbaum (1973, chap. 8) have said about entropy and the second law of thermodynamics. Consider the simplest sort of example. Suppose we have a box with two compartments that are connected by a window that can be opened or closed. The box contains equal numbers of nitrogen (N_2) and oxygen (O_2) molecules. The window is open, and all of the N_2 molecules are in the left-hand compartment, while all of the O_2 molecules are in the right-hand compartment. Suppose that there are two molecules of each type. If they are distributed randomly, there is a probability of $2^{-4} = 1/16$ that they would be segregated in just that way—a somewhat improbable coincidence.[17] If there are five molecules of each type, the chance of finding all of the N_2 molecules in the left compartment and all of the O_2 molecules in the right is a bit less than 1/1000—fairly improbable. If the box contains fifty molecules of each type, the probability of the same sort of segregation would equal $2^{-100} \simeq 10^{-30}$—extremely improbable. If the box contains Avogadro's number of molecules—forget it! In a case of this sort, we would conclude without hesitation that the system had been prepared by closing the window that separates the two compartments, and by filling each compartment separately with its respective gas. The window must have been opened just prior to our examination of the box. What would be a hopelessly improbable coincidence if attributed to chance is explained straightforwardly on the supposition that separate supplies of each of the gases were available beforehand. The explanation depends upon an antecedent state of the world that displays de facto orderliness.

Reichenbach generalized this point in his *hypothesis of the branch structure* (1956, sec. 16). It articulates the manner in which new sorts of order arise from preexisting states of order. In the thermodynamic context, we say that low entropy states (highly ordered states) do not emerge spontaneously in isolated systems, but, rather, they are produced through the exploitation of the available energy in the immediate environment. Given

[17] Strictly speaking, each of the probabilities mentioned in this example should be doubled, for a distribution consisting of all O_2 molecules in the left and all N_2 molecules in the right would be just as remarkable a form of segregation as that considered in the text. However, it is obvious that a factor of two makes no real difference to the example.

the fundamentality and ubiquity of entropy considerations, the foregoing parallel suggests that the conjunctive fork also has basic physical significance. If we wonder about the original source of order in the world, which makes possible both the kind of order we find in systems in states of low entropy and the kind of order that we get from conjunctive forks, we must ask the cosmologist how and why the universe evolved into a state characterized by vast supplies of available energy. It does not seem plausible to suppose that order can emerge except from de facto prior order.

In dealing with the interactive fork, I defined it in terms of the special case in which two processes come into an intersection and two processes emerge from it. The space-time diagram has the shape of an x. It does not really matter whether, strictly speaking, the processes that come out of the intersection are the same as the processes that entered. In the case of Compton scattering, for instance, it does not matter whether we say that the incident photon was scattered, with a change in frequency and a loss of energy, or say that one photon impinged upon the electron and another photon emerged. With trivial revisions, the principle CI can accommodate either description.

There are, of course, many other forms that interactions may exhibit. In the Annie Oakley example, two processes (the bullet and the knife-edge) enter the interaction, but three processes (the two bullet fragments and the knife-edge) emerge. We must be prepared to deal with even more complicated cases, but that should present no difficulty, for the basic features of causal interactions can be seen quite easily in terms of the x-type.

Two simpler kinds of interactions deserve at least brief mention. The first of these consists of two processes that come together and fuse into a single outgoing process. Because of the shape of the space-time diagram, I shall call this a λ-type interaction. As one example, consider a snake and a mouse as two distinct processes that merge into one as the snake ingests and digests the mouse. A hydrogen atom, which absorbs a photon and then exists for a time in an excited state, furnishes another.

The other simple interaction involves a single process that bifurcates into two processes. The shape of the space-time diagram suggests that we designate it a y-type interaction. An amoeba that divides to form two daughter amoebas illustrates this sort of interaction. A hydrogen atom in an excited state, which emits a photon in decaying to the ground state, provides another instance.

Since a large number of fundamental physical interactions are of the y-type or the λ-type (see, e.g., Feynman, 1962, or Davies, 1979), there would appear to be a significant advantage in defining interactive forks in terms of these configurations, instead of the x-type. Unfortunately, I have not seen how this can be accomplished, for it seems essential to have two processes going in and two processes coming out in order to exploit the idea of mutual modification. I would be more than pleased if someone could show how to explicate the concept of causal interaction in terms of these simpler types.

Concluding Remarks

There has been considerable controversy since Hume's time regarding the question of whether causes must precede their effects, or whether causes and effects might be simultaneous with each other. It seems to me that the foregoing discussion provides a reasonable resolution of this controversy. If we are talking about the typical cause-effect situation, which I characterized previously in terms of a causal process joining two distinct interactions, then we are dealing with cases in which the cause must precede the effect, for causal propagation over a finite time interval is an essential feature of cases of this type. If, however, we are dealing simply with a causal interaction—an intersection of two or more processes that produces lasting changes in each of them—then we have simultaneity, since each process intersects the other at the same time. Thus it is the intersection of the white light pulse with the red filter that produces the red light, and the light becomes red at the very time of its passage through the filter. Basically, propagation involves lapse of time, while interaction exhibits the relation of simultaneity.

Another traditional dispute has centered upon the question of whether statements about causal relations pertain to individual events, or whether they hold properly only with respect to classes of events. Again, I believe, the foregoing account furnishes a straightforward answer. I have argued that causal processes, in many instances, constitute the causal connections between cause and effect. A causal process is an individual entity, and such entities transmit causal influence. An individual process can sustain a causal connection between an individual cause and an individual effect. Statements about such relations need not be construed as disguised generalizations. At the same time, it should be noted, we have used statistical relations to characterize conjunctive forks. Thus, strictly speaking, when we invoke something like the principle of the common cause, we are implicitly making assertions involving statistical generalizations. Causal relations, it seems to me, have both particular and general aspects.

Throughout the discussion of causality, in this chapter and the preceding one, I have laid particular stress upon the role of causal processes, and I have even suggested the abandonment of the so-called event ontology. It might be asked whether it would not be possible to carry through the same analysis, within the framework of an event ontology, by considering processes as continuous series of events. I see no reason for supposing that this program could not be carried through, but I would be inclined to ask why we should bother to do so. One important source of difficulty for Hume, if I understand him, is that he tried to account for causal connections between noncontiguous events by interpolating intervening events. This approach seemed only to raise precisely the same questions about causal connections between events, for one had to ask how the causal influence is transmitted from one intervening event to another along the chain. As I argued in chapter 5, the difficulty can be circumvented if we look to processes to provide the causal connections. Focusing upon processes rather than events has, in my opinion, enormous heuristic (if not systematic) value. As John Venn said in 1866, "Substitute for the time honoured 'chain of causation,' so often introduced into discussions upon this subject, the phrase a 'rope of causation,' and see what a very different aspect the question will wear" (Venn, 1866, p. 320).

7 | Probabilistic Causality

HARDLY A DAY passes without a report in the morning newspaper about some causal claim. We have been informed in recent years that saccharin causes bladder cancer in laboratory animals. This experimental results lends some weight to the supposition that it may cause the same disease in humans. A suspicion has been voiced that substances used by woodworkers—epoxies or resins perhaps—cause cancer of the rectum or colon. There has been strong evidence for many years that exposure to intense radiation, such as is present in the vicinity of the explosion of a nuclear bomb, causes leukemia. Twenty years ago, I gave up cigarette smoking (it wasn't easy) because I believed it causes lung cancer. Subsequent studies indicate that it causes heart attacks and many other diseases as well. During the 1960s, evidence that indicated that the use of oral contraceptives causes thrombosis began to emerge. Countless other examples of the same sort could easily be added to the foregoing list.

The main thing that all of these examples have in common—apart from the fact that they all involve medical problems—is their striking failure to fit the Humean picture of constant conjunction. In the original saccharin studies, 100 rats were randomly selected from a population of 200 and were given high dosages of saccharin in their diets. The remaining 100 rats were treated in the same way, except that they were not given saccharin. In the experimental group, 14% of the rats developed bladder cancer, while only 2% of the control group did so (see Giere, 1979, pp. 247–256). The evidence that led to the suspicion that woodworkers are exposed to carcinogens was the fact that among the more than sixteen hundred workers in the General Motors woodshop during a ten-year period, fourteen (less than 1%) developed cancer of the rectum or colon. The normal rate for a sample of this size would be six cases. Of the 2,235 soldiers who witnessed an atomic blast at close range in operation Smoky in 1957, eight subsequently contracted leukemia; nevertheless, in the opinion of the medical investigator who examined the evidence, there was "no doubt whatever" that the radiation had caused the leukemia (see Salmon, 1978, pp. 688–689 and note 15). Even among individuals who are heavy cigarette smokers for protracted periods, lung cancer is by no means an inevitable consequence. As for the thrombosis caused by the use of oral contraceptives, there are about fifteen cases per million users, but this is about seven times as great as the rate in the population at large (Giere, 1979, pp. 272–277). The crucial factor in all of these examples is that the incidence of the pathological condition is significantly different in the presence of the alleged cause than in its absence.

THE SUFFICIENCY/NECESSITY VIEW

Hume considered his example of the colliding billiard balls "as perfect an instance of the relation of cause and effect as any which we know," and he commented in general that "every object like the cause produces always some object like the effect" (Abstract of the *Treatise*). Other standard examples are that decapitation causes death and that an electric spark passed through a mixture of hydrogen and oxygen causes the formation of water. In each of these cases, the cause is seen as a sufficient condition of the effect. It is, of course, necessary to exclude vacuous sufficient conditions. We do not want to say that being born is the cause of death, but with some appropriate restrictions the problem can, no doubt, be handled. Mackie's appeal to INUS (*insufficient nonredundant* part of *unnecessary* but *sufficient*) conditions (1974, p. 62) is one way to deal with it. Let us use the phrase "sufficiency/necessity view" to cover any analysis of causality that demands constant conjunctions in the form of sufficient conditions, necessary conditions, or any combination of the two.

The reader who has absorbed the main thrust of chapter 2 will recognize immediately that the basic feature of the examples cited in the first paragraph is that all of them involve relations of statistical relevance. The ingestion of saccharin by laboratory rats (and humans, perhaps) is positively relevant to the occurrence of bladder cancer. Exposure to epoxies, resins, or other substances in the woodshop appears to be positively relevant to the incidence of cancer of the rectum or colon. Cigarette smoking is positively relevant to affliction with lung cancer and numerous other diseases. The use of oral contraceptives is positively relevant to the occurrence of thrombosis. Proximity of the explosion of a nuclear bomb is positively relevant to the onset of leukemia. Whatever the basic *nature* of the causal relation itself, when we want to test for the presence of a cause-effect relation we seek *evidence*, not in the form of a constant conjunction, but, rather, in the form of a statistical relevance relationship. This point was brought out in chapter 2, when we discussed the question of the efficacy of vitamin C in the prevention or reduction of severity of the common cold. It continues to be a critical question in the evaluation of various forms of psychotherapy.

Our natural response to these observations might be a claim that although statistical relevance relations provide *evidence* for causal relations, causal relations are not appropriately *analyzable* in terms of statistical relevance relations. Take the case of thrombosis. The fact that use of oral contraceptives increases the probability of thrombosis shows, it might be said, that the oral contraceptive is a causal factor in the occurrence of thrombosis, but not *the* cause. Recall Mackie's appeal to INUS conditions. It is the rule rather than the exception that several different sets of conditions are each sufficient to produce an effect (in chapter 5 we discussed a house burning down), and that each of these sufficient conditions consists of a complex set of necessary components. When we say that the use of oral contraceptives is a causal factor in the occurrence of thrombosis, this is to be understood to mean that it is one factor in a combination of factors that, taken together, are sufficient to produce thrombosis. There might, for example, be certain hereditary dispositions to circulatory disease that also contribute to the occurrence of thrombosis in these cases. Additional causal factors—not yet known to medical science, perhaps—may also contribute. We cannot rightly claim to have found the cause of thrombosis—so the argument might continue—until we have identified that set of factors that is jointly sufficient to produce thrombosis. This is the kind of response to the thrombosis example that is to be expected from those who hold a sufficiency/necessity view. Parallel sorts of comments would apply to the other examples mentioned in the first paragraph of this chapter.

We can, I believe, use the framework developed in connection with the S-R basis of scientific explanation to interpret the foregoing approach. Suppose we start with some reference class, say, the class of all women of child-bearing age. (In this context, we won't worry about poor old John Jones who took his wife's birth control pills, even though he might have developed thrombosis.) Since the occurrence of thrombosis is the object of our attention, we take as our attribute class the class of women who suffer thrombosis; the explanandum-partition is simply {develops thrombosis, does not develop thrombosis}. We find that we can make a partition of our original reference class, in terms of the use of oral contraceptives, that is relevant to thrombosis. The cells in this partition are patently inhomogeneous with respect to thrombosis; for example, pregnancy is statistically relevant to thrombosis, as are numerous other factors, no doubt. If we could find all of these additional relevant factors, we could then provide a partition of the reference class that would be homogeneous with respect to thrombosis. Now, according to the analysis of causality in terms of INUS conditions, or any other version of a sufficiency/necessity theory, we cannot claim to have found the cause(s) of a given instance of thrombosis unless that case belongs to a cell in the homogeneous partition in which *every* member has the attribute of suffering a thrombosis. Since we are dealing with finite classes, it is equivalent to say that the attribute must have a probability equal to one in that cell. If the probability falls short of unity, or if it is impossible in principle to find an objectively homogeneous class, that particular thrombosis has no cause.

I do not deny that an analysis of this sort can be carried through consistently, but it seems misdirected for two reasons. First, I believe that there are clear cases of cause-effect relations that defy any analysis of that kind. Second, even in situations that may ultimately permit a sufficiency/necessity treatment, that approach involves a great deal of unnecessary metaphysical baggage. I object, not primarily on the ground that the baggage is metaphysical, but, rather, because it is unnecessary.

Let me begin by offering three examples that seem to me to exhibit cause-effect relations but that are in principle not amenable to an analysis in terms of sufficient and/or necessary conditions.

(1) The fact that an ice cube was placed in tepid water caused it to melt. According to the classical kinetic theory, there is a minute but nonvanishing probability that heat may flow from a colder body to a hotter body; consequently, it is possible, though highly improbable, that the water might become warmer as a result of heat given up by the ice cube, while the ice cube would become even colder and remain in a frozen state. Even if one takes the laws of classical mechanics, on which the classical kinetic theory is grounded, to be deterministic, there can be no guarantee that the initial conditions in the water and the ice will not be such as to prevent the melting of the ice. Therefore, being placed in tepid water is not a sufficient condition, and it is obviously not necessary.

(2) The lift that enables airplanes to take off is produced by a pressure difference between the upper and lower surfaces of the wing. Given that the wing has the proper shape of an airfoil, the Bernoulli principle accounts for the crucial pressure difference. In this case, as in the preceding one, the basic theory is statistical, and the result has a probability that falls just a tiny bit short of unity under the given conditions. Inasmuch as an airplane might be taken aloft piggyback by a Boeing 747—in the same fashion as the space shuttle—Bernoulli lift by the wings of the airplane in question is not necessary.

(3) In one type of laser, a large number of atoms are in a particular excited state. If photons of a suitable frequency impinge upon these atoms, they will decay rapidly to the ground state, and a burst of radiation will occur. This is the phenomenon of stimulated emission; the acronym LASER stands for "light amplification by stimulated emission of radiation." The cause of the burst of radiation is the incident light, which thus stimulates the atoms to emit radiation as they make a transition from the excited state to the ground state. The incident radiation is, however, neither a necessary nor a sufficient condition for the burst of radiation. Each atom has a certain

probability of remaining in the excited state for a given length of time; if incident radiation of the appropriate frequency impinges, the probability of an earlier transition is greatly increased. It is therefore possible for all of the atoms to make a rapid transition to the ground state spontaneously, without the help of incident radiation, though this occurrence would be extremely improbable. Similarly, it is possible—though, again, highly improbable—for the atoms to remain in the excited state, even in the presence of light of the frequency appropriate to stimulate emission. In this example, in contrast to the preceding two, the fundamental physical theory is generally considered to be *irreducibly* statistical.

In all three of the foregoing examples, it seems to me quite unreasonable to deny that cause-effect relations are present. Many other cases, similar in principle, can easily be devised. I find in such examples compelling (though *not* absolutely incontrovertible) evidence that cause-effect relations of an ineluctably statistical sort are present in our universe.

Some readers are likely to feel that I have been indulging in technical quibbles—maintaining that, for all practical purposes, examples (1)–(3) do involve sufficient and/or necessary conditions. For those who react in this way, I should like to consider several commonplace examples. These will be useful in formulating my second objection to the sufficiency/necessity thesis.

(4) An example, already mentioned in the preceding chapter, will serve here. It is the case of the window broken by the baseball. It seems altogether plausible to suppose that windowpanes of the same size, shape, and thickness shatter in 95% of all instances in which they are struck by baseballs of the same size traveling at the same velocity. The fact that breakage does not occur in every case does not constitute an adequate ground for denying that the impact of *this* baseball caused *this* window to break. Being struck by a baseball is clearly not necessary, for the window might have been shattered by an earthquake.

(5) An automobile accident may be due to fatigue on the part of the driver. This driver, and other similar drivers, do not always fall asleep at the wheel when they are tired. When a driver does fall asleep, it does not always lead to an accident. On some occasions, when the driver nods, the car veers toward the lane of oncoming traffic; on other occasions, it goes toward the shoulder. Again, it seems to me, we need not deny the causal relation just because fatigue does not always result in a collision. Furthermore, since similar accidents are sometimes due to drunkenness, fatigue is not necessary.

(6) After dining on well-seasoned Mexican cuisine, I sometimes experience a bit of gastric distress. On such occasions, I have no hesitation

in saying that the food caused the discomfort, even though it occurs neither always nor only after the consumption of spicy meals.

It is easy to imagine the sort of reply a sufficiency/necessity theorist would be likely to make to examples like (4). If we take into account the precise momentum (both magnitude and direction) of the baseball and the precise microstructure of the glass in the windowpane, we will be told, then it is completely determined that the pane will shatter when struck by the ball. We must, of course, take other relevant conditions into account—for example, the hardness of the ball, whether the stitching makes direct contact with the glass, the temperature of the glass—and many others.

I have three rejoinders to offer to this line of argument. First, if the proponent of the sufficiency/necessity view is one who took exception to examples (1)–(3) on the grounds of technical quibbling, then I'd be inclined to ask *tu quoque* who is dabbling in technicalities now? If it is legitimate to appeal to precise values of momentum in example (4), for instance, why should it be illegitimate to take account of precise values of probabilities in the first three examples?

Second, if we are going to deal in technicalities in order to support one position on causality or the other, we should at least have some reasonably adequate scientific basis for our claims. I suggest that my remarks about example (3), which concerns the laser, have a firm basis in a well-supported scientific theory. In contrast, the appeal of the sufficiency/necessity theorist to the microstructure of glass seems not well-founded at all. The main support for the deterministic analysis of causality seems to come from a Laplacian view of classical physics. One of the most notorious shortcomings of classical physics was its inability to provide even a halfway satisfactory account of the structure of matter. Although enormous progress has subsequently been made on this front, the structure of glass is still not well understood (see Phillips, 1982). Thus, while the statistical analysis of example (3) has reasonable scientific support, the sufficiency/necessity analysis of example (4) has no basis beyond sheer speculation.

Third, it seems altogether unnecessary to burden our common-sense concept of causality with the dubious metaphysical thesis of determinism. Suppes puts the point well when, in commenting upon examples similar to (4)–(6), he remarks:

It is easy to manufacture a large number of additional examples of ordinary causal language, which express causal relationships that are evidently probabilistic in character. One of the main reasons for this probabilistic character is the open-textured nature of analysis of events as expressed in ordinary language. The completeness and closure conditions so naturally a part of classical physics are not at all a part of

> ordinary talk. Thus in describing a causal relation in ordinary circum-
> stances, we do not explicitly state the boundary conditions or the lim-
> itations on the interaction between the events in question and other events
> that are not mentioned. (1970, p. 8)

I cannot think of any reason to suppose that ordinary causal talk would
dissolve into nonsense if Laplacian determinism turned out to be false. I
shall therefore proceed on the supposition that probabilistic causality is a
coherent and important philosophical concept.

In advocating the notion of probabilistic causality, neither Suppes nor
I intend to deny that there are sufficient causes; indeed, Suppes explicitly
introduces that concept into his theory (1970, p. 34, def. 9). On our view,
sufficient causes constitute a limiting case of probabilistic causes. On the
sufficiency/necessity view, which we reject, this limiting case includes all
bona fide cause-effect relations. This latter approach to causality seems
needlessly restrictive.

STATISTICAL RELEVANCE AND PROBABILISTIC CAUSALITY

In the discussion of the statistical-relevance (S-R) basis of explanation,
it will be recalled, we recognized the possibility that an event-to-be-explained might belong to a cell in the homogeneous partition of the original
reference class in which the probability in question equals one. These are
the kinds of instances cherished by those who hold the view that deductive-
nomological (D-N) explanations are the only legitimate scientific explanations. This doctrine characterizes those who, like von Wright or Harré
and Madden, adopt the strict modal conception of scientific explanation.[1]
Many proponents of the epistemic conception—most notably Hempel—
admit that there are scientifically acceptable statistical explanations. Because Hempel explicitly rejected the thesis that explanations must be causal,
the admission of statistical explanations did not force him to deal with the
concept of probabilistic causality. Prior to 1976, Hempel maintained that
explanations that fit his inductive-statistical (I-S) pattern must confer high
probabilities upon their explanandum-events. Later, chiefly in response to
Jeffrey (1969), Hempel admitted the possibility of explanations that do not
render the events-to-be-explained highly probable in relation to their premises. On Hempel's view, it seems to me, we can regard D-N explanations
as a special limiting case of scientific explanations in general. Those who
adopt the strict modal conception, in contrast, can admit as legitimate only

---

[1] These remarks would not apply to Mellor, who—although he subscribes to a modal
conception—allows for probabilistic explanation by appealing to degrees of necessity. I have
explained in chapter 4 why I think his theory should be regarded as a causal theory.

those explanations that constitute merely a limiting case for many proponents of the epistemic conception.

The classic version of the epistemic conception of scientific explanation is the inferential version. It is easy to see how D-N explanations fit within that framework, for deductive arguments seem to do exactly what we want them to do. Moreover, it appears (at least superficially) not too difficult to accommodate I-S explanations with high associated probabilities, for strong inductive arguments are quite widely taken to be acceptable. It is far more difficult to see how statistical explanations with low probabilities are to be accommodated within the epistemic conception, but if one is willing—as Hempel appears to be—to adopt some sort of Carnapian inductive logic that has no inductive rules of acceptance, then one can adopt the rather Pickwickian sense of "inference" proposed by Carnap, and still hold onto an inferential conception of scientific explanation (see chapter 4).

One significant feature of the inferential version of the epistemic approach, as just sketched, is that it accords no role whatever to relations of statistical relevance. In fact, to the best of my knowledge, the only place in which Hempel employs the expression "statistically relevant" is in his (1968), where he defines it explicitly as a certain type of simple probability relation—$P(G,F_1) = r$—rather than a relation between two probability values, as in the standard meaning of the term. Thus, it seems to me, he has not appreciated what I take to be *the* most fundamental objection to his version of the epistemic approach. Hempel has focused exclusively upon the absolute size of the probability associated with the explanandum-event in relation to the explanans, and he has entirely overlooked the need to make a comparison between two probabilities—a prior probability and a posterior probability—as spelled out in chapter 2.

When Hempel published his first systematic account of I-S explanation (1962), I was immediately struck with its failure to recognize the importance of statistical relevance relations. I made this point in my first paper on the subject (1965), and have reiterated it many times subsequently. For quite some time, I was genuinely puzzled by the failure of Hempel, as well as other advocates of the inferential approach, to see the crucial importance of relevance relations. A possible explanation now occurs to me. My initial response to Hempel's I-S model was to propose an alternative model of statistical explanation. As this S-R model was first presented (Salmon et al., 1971), it incorporated no causal features; in this respect, it was on a par with Hempel's I-S model in being purely statistical and acausal. It seemed obvious at the time that statistical relevance relations had some sort of explanatory power in and of themselves. As I have said repeatedly throughout this book, that view now appears to be utterly mistaken. The explanatory significance of statistical relevance relations is indirect. Their fundamental import lies in the fact, emphasized at the beginning of this chapter, that they constitute evidence for causal relations. However, for those who share Hempel's view that causal relations play no essential role in scientific explanation, that fact will carry little if any weight. For those who adhere to the ontic conception of scientific explanation, the relation between statistical relevance and causality will confer special importance upon statistical relevance relations.

CAUSALITY AND POSITIVE RELEVANCE

The relation of positive statistical relevance plays a fundamental role in every theory of probabilistic causality with which I am acquainted. As formulas (3) and (4) of chapter 6 show, positive statistical relevance of cause to effect is a condition used to characterize conjunctive forks. It plays an equally important part in Reichenbach's definition of causal betweenness (1956, p. 190). In Suppes's probabilistic theory of causality, an earlier occurrence is a *prima facie cause* of a later occurrence if the earlier event is positively relevant to the later one (1970, p. 12). A prima facie cause is *spurious* if it is screened off—that is, rendered irrelevant—by a still earlier event (1970, pp. 23–25). In cases of spurious causation—for example, the relationship between the falling barometric reading and the subsequent storm—the positive correlation is explained in terms of an earlier common cause. Spurious causation is closely related to Reichenbach's conjunctive forks. Causes that are not spurious are *genuine* (1970, p. 24). Suppes does admit negative causes; they are defined in terms of negative rather than positive statistical relevance (1970, p. 44). All other types of causes mentioned by Suppes—for example, direct causes, indirect causes, supplementary causes, sufficient causes—presuppose positive statistical relevance. In I. J. Good's causal calculus (Good, 1961–1962), a measure of "the tendency of F to cause E," which bears a close relation to Suppes's relation of genuine causation, holds a central place. I have discussed these three theories in detail in Salmon (1980).[2]

Plausible as Suppes's treatment of genuine causation may seem at first

[2] Additional theories of probabilistic causality can be found in (Fetzer and Nute, 1979), (Sayre, 1977), and (Tuomela, 1977). Fetzer and Nute employ a possible-worlds approach; I have explained in (Salmon, 1976, pp. x–xii) why I believe that kind of approach to a wide variety of philosophical problems is inadvisable. Sayre develops his theory of probabilistic causality on the basis of certain communication-theoretic concepts. As I remarked in chapter 4, this approach has considerable interest. Tuomela uses the concept of inductive explanation to explicate probabilistic causality. It should be clear that, from my standpoint, this approach seems to go in the wrong direction.

blush, it does appear open to two major objections. In the first place, the definition is too broad; it admits, as cases of genuine causation, examples that clearly should not qualify. In the second place, it appears to be too narrow, for it seems to exclude cases in which genuine causation is present. I am much more confident of the first objection than I am of the second.[3]

The first difficulty arises in connection with interactive forks, which are discussed at some length in the preceding chapter. Consider, for example, the case of the inexperienced pool player. The balls were situated on the table in such a way, it will be recalled, that the player had a 50-50 chance of sinking the 8-ball in one of the far corner pockets, but if he did so, the cue ball was almost certain to drop into the other far corner pocket for a 'scratch.' Under these circumstances, the 8-ball could be expected to fall into the pocket just a little bit earlier than the cue ball would fall into its pocket. The dropping of the 8-ball is thus a prima facie cause of the scratch. As noted in our earlier discussion of this example, the collision of the cue ball with the 8-ball does not statistically screen these events from one another. Given the shot by the novice—including the fact that the cue ball actually makes contact with the 8-ball—there is a probability of 1/2 that the cue ball will go into the pocket. Given the shot by the novice, and the fact that the 8-ball has dropped, the probability that the cue ball will go into the pocket is nearly one. As we saw in the previous chapter, this kind of failure of screening off is typical of interactive forks. Thus, in this case, as well as many other examples of interactive forks, one effect qualifies under Suppes's definition as a genuine cause of the other.

The second difficulty is illustrated by an example due to Deborah Rosen; it is explicitly discussed by Suppes (1970, p. 41). Although Suppes refers to it as the problem of "improbable consequences," I think it is more appropriately described as *the problem of negative relevance*. Let us consider a slightly modified version of Rosen's original example. Suppose that a golfer of only moderate skill tees off on a par two hole at a pony-golf course. The shot is badly sliced, but by the sheerest accident the ball hits a branch of a tree near the green and drops into the hole for a spectacular hole-in-one. Let A stand for the tee stroke, let D stand for the collision of the ball with the branch, and let E stand for the ball dropping into the hole. The problem is this. The events A, D, E seem to be links in a causal chain, but given A, we must admit that D is negatively relevant to E— that is, $P(E|A) > P(E|A.D)$. The probability that this player will make a hole-in-one on this particular hole is surely greater than the probability that she will make a hole-in-one by bouncing the ball off of a tree branch. The same problem arises in connection with Reichenbach's definition of

---

[3] A penetrating critique of Suppes's theory can be found in (Otte, 1981).

causal betweenness (1956, p. 190), for he requires that $P(E|A.D) > P(E|A)$ if D is causally between A and E.

Let us consider another example in which a cause seems to be negatively relevant to its effect. At the beginning of this chapter, I mentioned the fact that some investigators believe there is evidence to support the view that consumption of oral contraceptives causes thrombosis. Germund Hesslow (1976) uses this example to criticize Suppes's theory. There is reason to believe, as Hesslow points out, that pregnancy also causes thrombosis, and it is at least possible that the probability of a pregnant woman suffering thrombosis is greater than the probability (which is *very* small) that a woman who takes birth control pills will suffer thrombosis. Indeed, it is quite possible that the probability of thrombosis in the class of women who use oral contraceptives is lower than the probability of thrombosis in the class of all women of child-bearing age, since oral contraceptives prevent a more potent cause of thrombosis (pregnancy) from being present.

There is, of course, a straightforward answer to Hesslow's criticism. If we take the class of women of child-bearing age and partition it into the class of women who become pregnant and those who do not, then we may find that the incidence of thrombosis is significantly higher in the former subclass than in the latter. This may constitute grounds for a suspicion that pregnancy probabilistically causes thrombosis, and further investigation may confirm this suspicion. Let us therefore assume, for purposes of this example, that pregnancy is one cause of thrombosis. We now look at the other subclass—namely, women who do not become pregnant. We partition this class into the subclass of women who take birth control pills and those who do not. If, as the evidence seems to indicate, the incidence of thrombosis is greater in the former subclass (those who use oral contraceptives) than in the latter, then we have grounds for suspicion that the use of oral contraceptives is another probabilistic cause of thrombosis. The use of oral contraceptives is positively relevant to thrombosis when we pick the appropriate initial reference class.

Rosen's example of the near-miraculous hole-in-one is not quite as easy to deal with. Two distinct ways of handling it have been proposed. The first might be called *the method of more precise specification of events*. Following this approach, one might argue that if we specify precisely the motion of the golf ball, as well as the position of the tree branch relative to the hole, then it may turn out that striking the branch *in this particular way* is positively relevant to the ball dropping into the hole.

I find this type of answer thoroughly unconvincing. Whether the ball will drop into the hole is extremely sensitive to minute changes in the conditions of the collision. For example, the outcome of the collision depends upon the angular momentum as well as the linear momentum of the ball. A slight difference in the spin of the ball that has been sliced—and slicing imparts spin—could make the difference between the hole-in-one and no hole-in-one. In addition, the outcome of the collision depends critically upon the position of the branch. Assuming, quite plausibly, that the surface of the branch is rather uneven, a small breeze could change the position of the branch enough to make the difference. Moreover, since the collision with the branch is obviously not an elastic collision, an unrealistically detailed description of the surface texture of the branch would be required to yield even a reasonable probability for the hole-in-one. The claim that positive statistical relevance can always be restored in examples of this sort by providing a sufficiently precise description of the events and conditions involved seems to me no better than the claim that the shattering of windows by baseballs can be rendered deterministic if sufficiently precise information is supplied.

The second way of handling Rosen's example might be called *the method of interpolated causal links*. Using this approach, one might admit that the probability of getting a hole-in-one by bouncing the ball off a tree branch is smaller than the probability of getting a hole-in-one regardless of how it comes about—in particular, simply by making an excellent shot from the tee. However, once it has happened that the drive is sliced, and is traveling in a direction toward the rough rather than the hole, then the probability of a hole-in-one is enhanced by the collision with the branch. If the ball does not make contact with the branch (or some other suitably placed object), the probability of the hole-in-one is for all practical purposes zero. If the ball hits the branch, the probability, though small, is not entirely negligible. The difficulty with this approach is that it merely shifts the problem. If we allow that the ball traveling through the air, after having been sliced, qualifies as an event C, then that event will be negatively relevant to the hole-in-one. The probability $P(E|A)$ that the player will make a hole-in-one, given that she tees off, is certainly larger than the probability $P(E|A.C)$ that she will make a hole-in-one, given that she tees off and slices badly. Nevertheless, C is taken as a link in the causal chain, and it is causally between the shot from the tee and the ball dropping into the hole.

It would be a serious mistake to suppose that the problem of negative relevance arises only in rare and outlandish cases; in fact, it is a common phenomenon. Whenever a given type of outcome can be probabilistically produced in two or more different ways—where the probabilities associated with the alternatives differ—this problem will arise. In the game of craps, for example, the player may win by throwing 7 or 11 on the first toss, or by throwing another number (not including 2, 3, or 12) and by 'making the point' by throwing that number again before getting 7. As the player begins the game, the chance of winning in one or another of the alternative ways is just slightly less than 1/2. If 4 is thrown on the first toss, the chance of winning drops to 1/3; therefore, throwing 4 is negatively relevant to winning. If, nevertheless, the player wins by making that point, the toss of the 4 is an event in the causal chain.[4]

To reinforce the commonplace character of the problem of negative relevance, consider another example. Suppose that two and only two candidates are running for the office of governor of a particular state, and that both candidates advocate a major bridge reconstruction project. Assume that candidate A has a 2/5 chance of winning, while candidate B has a chance of 3/5. If candidate A is elected, there is a probability of 1/10 that she will be willing and able to persuade the legislature to pass the appropriation bill that will eventuate in the reconstruction project; if candidate B is elected, there is a probability of 1/2 that he will be able to accomplish that end. Given these values, the probability of the bridge reconstruction is 0.34. Suppose, however, that A is elected, that she endeavors to get the appropriation bill enacted, and that her efforts succeed. The fact that she was elected is negatively relevant to the bridge reconstruction, for it reduces the probability from 0.34 to 0.1; nevertheless, it is part of the causal chain.

The problem of negative relevance is both ubiquitous and profound—ubiquitous because there are usually "many ways to skin a cat," and profound because it cuts against a powerful intuition that is built into the foundations of virtually every probabilistic theory of causality. To the extent that it has been recognized in the literature, it has almost always been handled by *the method of more precise specification of events* or *the method of interpolated causal links*, or a combination of the two. I have indicated above why I think that none of these approaches is satisfactory. There is, however, another approach, which might be designated *the method of successive reconditionalization*, which may be somewhat more successful (see Salmon, 1980, pp. 68–69).

We can construct a simple two-stage game to illustrate this method. To play this game, the player tosses a regular tetrahedron whose sides are marked 1, 2, 3, 4, respectively. If the tetrahedron comes to rest on any side other than 4—that is, if side 4 shows after the tetrahedron comes to rest—the player draws from a deck containing sixteen cards, twelve of which are red and four of which are black (hereinafter, the red deck). If side 4 does not show, he draws from a deck containing four red and twelve black cards (hereinafter, the black deck). The object of the game is to draw a red card. A simple calculation shows that the probability of drawing

---

[4] See (Copi, 1978, pp. 522–523) for the rules of this game and for values of some of the basic probabilities.

a red card if one plays this game is 10/16. Suppose that this game is one among many different games set up in a particular gambling house. A patron in this establishment may choose to play this game, whereupon he puts down a certain stake and tosses the tetrahedron. Consider a particular case in which a player enters the game (B) and tosses the tetrahedron. Side 4 shows, so he draws a card from the red deck (E). The draw results in a win, for he picks a red card (F). We have a causal chain $B \rightarrow E \rightarrow F$, where each event is positively relevant to its successor. Entering this game is positively relevant to getting the 4 to show, for that is the only way in which a player has access to the tetrahedron. Drawing a card from the red deck is positively relevant to drawing a red card, for the probability of drawing a red card from the red deck is 3/4, while the overall probability for getting a red card in the game is 10/16.

Consider, however, another instance in which the player enters this game, but when he tosses the tetrahedron, 4 does not show. He must draw from the black deck. Luck is with him and he draws a red card. In this chain of events, the successive links are not positively relevant to one another; in particular, drawing from the black deck is negatively relevant to getting a red card. The probability of getting a red card if he draws from the black deck is 1/4, which is less than the overall probability of getting a red card in this game. Nevertheless, in this instance, the draw from the black deck is the probabilistic cause of the winning draw.

The basic idea behind the method of successive reconditionalization is that once a particular event in the causal chain has occurred, it does not matter what other events might have happened, but did not. In the present example we want to say, roughly speaking, that once the tetrahedron has come to rest with side 4 on the bottom, drawing from the black deck *is* positively relevant to getting a red card, for that is the only way for a player to get a red card, given the outcome of the toss of the tetrahedron. The fact that a different outcome of the toss would have provided the player with a better chance of getting a red card is quite beside the point.

In order to make this approach more precise, let us look at the events that transpire in our fictitious game in greater detail. Using letters in a way that is consistent with preceding notation, we can designate the salient events in the game as follows:

A - A player in the casino makes a decision to play a particular game.
B - The player enters this game by putting down his stake.
C - The player tosses the tetrahedron used in our game.
D_1 - The tetrahedron comes to rest with side 4 showing.
D_2 - The tetrahedron comes to rest with side 4 on the bottom.
E_1 - The player draws from the red deck.

E_2 - The player draws from the black deck.

E_3 - The player draws from neither deck.

F - The player gets a red card and wins a prize.

Our causal chain can be symbolized $A \to B \to C \to D_2 \to E_2 \to F$. In order to ascertain the pertinent statistical relevance relations, we shall reconditionalize our probabilities on each event in the chain as it occurs. Let us take A, the class of decisions, as our universe; within this class, some decisions are decisions to enter the game we are considering. Given that the decision has been made to enter our game, the question concerning the nature of the other games available is irrelevant. It does not matter whether there are other games that employ tetrahedrons, or whether there are other games that involve the tossing of tetrahedrons that have a probability greater than 3/4 of showing side 4. Once the player in our game has made the toss on which 4 does not show, and is forced to draw from the black deck, it does not matter that there is another deck from which he would have drawn had the result of the toss been different.

Let us consider carefully the following four probability relations:

$$P_A(C|B) > P_A(C|\bar{B}) \tag{1}$$
$$P_B(D_2|C) > P_B(D_2|\bar{C}) \tag{2}$$
$$P_C(E_2|D_2) > P_C(E_2|\bar{D}_2) \tag{3}$$
$$P_{D_2}(F|E_2) > P_{D_2}(F|\bar{E}_2) \tag{4}$$

Inequality (1) is unproblematic. Since entering the game and placing a bet are necessary conditions for tossing this tetrahedron, the term on the right is zero. The term on the left will be large, but not necessarily unity, since it may happen occasionally, for example, that a police raid stops the game after the player has put down his stake, before he gets a chance to toss the tetrahedron. This fact is important to insure that the term on the right of (2) is well defined. Given this assumption, inequality (2) is also straightforward. Again, among those who have entered the game, the only way for the toss to result in side 4 being down is as a result of the toss of the tetrahedron; therefore, the term on the right vanishes, while, by the conditions stipulated for the game, the term on the left has the value 1/4. Inequality (3) presents no problems, for the term on the right is zero while the term on the left is almost unity. It is not strictly unity because, for example, a heart attack might prevent a player who has tossed the tetrahedron from drawing from any deck. The last inequality is a crucial one. If the two probabilities were conditioned on A rather than D_2, this inequality would not hold, for almost every player who does not draw from the black deck draws from the red deck. The term on the right would be larger, not smaller, than the term on the left. When, however, we take these proba-

bilities only over the restricted class of players who have made a toss of
the tetrahedron on which 4 did not show, the reverse inequality does not
hold. The major problem with this inequality is that the term on the right
would be meaningless if, within the class of players for whom the tetrahedron came to rest on side 4, everyone drew from the black deck. We
can plausibly maintain, however, that a player whose tetrahedron toss has
the unfavorable outcome may sometimes not draw from either deck, but
may (as just mentioned) have a heart attack, or may get up and walk away
in disgust. In such cases, the player fails to draw a red card, and thereby
forfeits the opportunity to win a prize; consequently, the term on the right
side is zero, while the term on the left is 1/4, thus satisfying (4). If this
analysis is correct, we have succeeded in showing that every event in our
causal chain is positively relevant to its immediate successor, provided we
reconditionalize appropriately at each step along the way.

Although I am inclined to think that the foregoing treatment of our
fictitious game is plausible, it may be felt that I have saved the analysis
by an inappropriately picky use of details—for example, in saving the
meaningfulness of the term on the right side of (4). For those who are
inclined to pose this sort of objection, I should like to present an alternative
formulation, by writing inequalities (1)–(4) in a slightly revised fashion:

$$P_A(C|B) \geqslant P_A(C) \tag{1'}$$
$$P_B(D_2|C) \geqslant P_B(D_2) \tag{2'}$$
$$P_C(E_2|D_2) \geqslant P_C(E_2) \tag{3'}$$
$$P_{D_2}(F|E_2) \geqslant P_{D_2}(F) \tag{4'}$$

where the equality is allowed only if the reference class for the probability
on the left is extensionally equal to the universe upon which the entire
probability is conditioned. With this formulation, we do not have to worry
if the class E_2 is the same as the class D_2.

Let us now try applying the method of successive reconditionalization
to Rosen's golf example. Let A be the class of tee shots by our player at
this particular hole, let B be a swing that produces a slice, let C be the
sliced ball traveling toward the tree, let D be the collision with the branch,
and let E be the dropping of the ball into the hole. The following relevance
relations hold:

$$P_A(C|B) > P_A(C|\bar{B}) \tag{5}$$
$$P_B(D|C) > P_B(D|\bar{C}) \tag{6}$$
$$P_C(E|D) > P_C(E|\bar{D}) \tag{7}$$

Inequality (5) says that the ball is more likely to be traveling in the direction
toward the tree if the tee shot is a slice than it is if the swing is a good
shot or a hook. Inequality (6) says that given the slice, the ball is more

likely to collide with the branch if it is headed in the direction of the tree than it is if it is going in another direction. Inequality (7) says that given the ball traveling toward the tree, it is more likely to go into the hole if it hits the branch than it is if it does not hit the branch. Each of these assertions seems manifestly correct.

The difference between this analysis of Rosen's example and the analysis by the method of interpolated causal links hinges upon the status of indirect causes. If we go back to the original formulation of the problem we have three events—the tee shot A, the collision with the branch D, and the dropping of the ball into the hole E. The problem was that D is not positively relevant to E, and hence, it is not even a prima facie cause of E. A fortiori, it cannot be a genuine cause of E in Suppes's theory. If we interpolate C—the ball traveling toward the tree—then, given C, D is positively relevant to E. In this case, however, C is negatively relevant to E, for our player has a better chance of getting a hole-in-one if she hits the ball well in the direction of the hole than she does if she slices it toward the tree. One might argue that C is not a direct cause of E, but we would surely want to consider it an indirect cause, since it is one of the events in the causal chain. On Suppes's account, unfortunately, it is not even a prima facie cause, so it cannot be an indirect cause (1970, p. 28). Similarly, on Reichenbach's theory, C is not causally between the teeing-off and the hole-in-one. When the method of successive reconditionalization is used, however, we are, in effect, dropping the requirement that indirect causes should be positively relevant to their ultimate effects—it is sufficient if they are positively relevant to their immediate effects, that is, their immediate successors in the causal chains. Indeed, following this approach, the concept of the indirect cause, as introduced by Suppes, becomes superfluous, since positive relevance is needed only step by step as we go along the causal chain.

In an attempt to see how far the requirement of positive relevance could be pushed, I introduced a fictitious example in atomic physics having the same statistical structure as the simple version of the tetrahedron/card example (1980, p. 65).

(7) Suppose that we have an atom in an excited state to which we shall refer as the 4th energy level. It may decay to the ground state (zeroeth level) in several different ways, some of which involve intermediate occupation of the 1st energy level. Let $P(m{\rightarrow}n)$ stand for the probability that an atom in the mth level will make a direct transition to the nth level. Assume that the probabilities have the following values:

$$P(4{\rightarrow}3) = 3/4 \quad P(3{\rightarrow}1) = 3/4$$
$$P(4{\rightarrow}2) = 1/4 \quad P(2{\rightarrow}1) = 1/4$$
$$P(3{\rightarrow}2) = 0$$

It follows that the probability that the atom will occupy the 1st energy level in the process of decaying to the ground state is 10/16; if, however, it occupies the 2nd level on its way down, then the probability of its occupying the 1st level is 1/4. Therefore, occupation of the 2nd level is negatively relevant to occupation of the 1st level. Nevertheless, if the atom goes from the 4th to the 2nd to the 1st level, that sequence constitutes a causal chain, in spite of the negative statistical relevance of the intermediate stage. Although this example is admittedly fictitious, one finds cases of this general sort in examining the term schemes of actual atoms.[5]

In view of the fact that we cannot, so to speak, 'track' the atom in its transitions from one energy level to another, it appears that there is no way, even in principle, of filling in intermediate events. Therefore, the method of *interpolating causal links* in inapplicable. Furthermore, it seems unlikely that the method of *more detailed specification of events* will work, for when we have specified the type of atom and its energy levels, there are no further facts that are relevant to the events in question. Moreover, in view of these facts, even the method of *successive reconditionalization* seems to fail. The pertinent inequality

$$P_4(1|2) < P_4(1|\bar{2})$$

goes in the wrong direction. If it were legitimate to conditionalize on nonoccupation of the 3rd energy level, we could write

$$P_{\bar{3}}(1|2) > P_{\bar{3}}(1|\bar{2}),$$

which has the correct sense, but nonoccupation of level 3 is *not* a physical event that precedes occupation of level 2. In discussing the golf ball, we could say that it was moving toward the branch rather than the hole after it had been driven off of the tee and before it struck the branch. In dealing with atomic transitions, however, there is no basis for saying that the atom is headed toward energy level 2 rather than energy level 3 after it has left the 4th level but before it reaches the 2nd. The sort of reconditionalization that seems to be required to save the principle of positive relevance appears to be blocked when we are dealing with examples to which the quantum theory applies.

Some doubt has been expressed by I. J. Good (1980, p. 303) as to the propriety of considering the foregoing atom a causal process under the conception I have been presenting. I do not find a serious problem here. An atom is a causal process, in the sense discussed in chapter 5, and it enters into causal interactions in the fashion treated in chapter 6. In the

[5] See, for example, the cover design on the well-known introductory text (Wichmann, 1967), which is taken from the term scheme for neutral thallium. The term scheme itself is given in Fig. 34A, p. 199.

first place, an atom can be marked, and it is capable of transmitting the mark. Consider a hydrogen atom. If the spin of the electron is parallel to that of the proton, then it can be made antiparallel by incident microwave radiation. Incident radiation of a different frequency can place the electron in a Bohr orbit above that of the ground state. Higher frequency radiation, as well as other mechanisms, can ionize the atom. An atom evidently satisfies the conditions imposed by the 'at-at' theory of causal influence. In the second place, each time the atom in the foregoing example makes a transition to a lower energy level, it participates in a y-type interaction, for each transition is accompanied by emission of radiation of the wavelength characteristic of that particular transition. Thus, it seem to me, we must give serious consideration to the idea that a probabilistic cause need not bear the relation of positive statistical relevance to its effect. One could say, of course, that the occurrence of the effect is more probable in the presence of a probabilistic cause than it is in the absence of any cause whatever (Good, 1980, p. 303). This observation amounts to the assertion that an event is more likely to occur in circumstances in which it is possible than it is in circumstances (the absence of any probabilistic cause) in which it is impossible. I do not find this answer especially illuminating.

CAUSAL PROCESSES AND PROPENSITIES

If positive statistical relevance is not the essential ingredient in a theory of probabilistic causality, then what is the fundamental notion? The answer, it seems to me, lies in the transmission of probabilistic causal influence. The point can best be seen, I think, by considering a number of concrete examples.

(8) If a toy sailboat is floating on a pond, and if a wave—produced by a rock dropped into the water nearby—approaches the boat, there is a certain probability that the boat will capsize when the wave reaches it. The probability will, of course, vary with the distance between the boat and the point of entry of the rock. For certain distances, the probability of capsizing will be unity and for others it will be zero for all practical purposes, but for intermediate distances, the probability can reasonably be supposed to have intermediate values (see Salmon, 1975b). The boat is a causal process, and the wave propagating over the surface of the pond is another. When they intersect, there is a certain probability for a certain type of interaction, and a certain probability that it will not occur.

(9) If the word "mother" is spoken to a subject in a psychological word-association experiment, there is a certain probability distribution that the subject will answer "earth," "child," "Mary," "father," another six-letter word beginning with "f," and so forth.

(10) When Rutherford's student Marsden bombarded gold foil with alpha particles, there was a certain probability that a given particle would pass through the foil without any noticeable effect, a certain probability that it would be deflected by a small angle, and a certain probability that it would bounce back in the direction whence it came.

(11) When Reggie Jackson's bat is swung at a pitched baseball, there is a certain probability that it will interact with the ball in a way which will send the ball clear out of the ball park.

(12) Given an amoeba floating in a pond, there is a certain probability that—within the next hour, for instance—it will undergo mitosis and divide into two daughter cells.

(13) Under suitable circumstances, an energetic gamma ray may vanish, producing an electron-positron pair.

(14) A given carbon 14 atom has a probability of 1/2 of emitting a negative electron and being transformed into nitrogen 14 within a period of 5,730 years.

(15) If a mouse is placed in a compartment with a boa constrictor, there is a certain probability that the boa constrictor will ingest, digest, and absorb the mouse into its system.

(16) If radiation of suitable frequency impinges upon a hydrogen atom, there is a certain probability that the radiation will be absorbed and the atom will exist for a time thereafter in an excited state.

The basic causal mechanism, in my opinion, is a causal process that carries with it probability distributions for various types of interactions. In chapter 6 I described three main kinds of causal interaction—x-type, y-type, and λ-type. The ten instances, just listed, exemplify these three types. The first four items (8)–(11) involve x-type interactions. In these cases, one causal process is propagating through space-time, and it carries probabilities for different sorts of interactions if it intersects with another process of a particular sort. It seems to me altogether appropriate to refer to these probabilities as *propensities*. Thus the sound "mother," impinging upon the auditory apparatus of a given subject, carries a set of propensities to elicit a set of responses. A given alpha particle, impinging upon a gold foil, has propensities of given magnitudes for no interaction, for small deflection, and for large deflection. Causal processes transmit energy, among other things, but they also transmit propensities for various kinds of interactions under various specifiable circumstances. As example (8) shows, the propensities may change as the process continues. The same point is illustrated by example (4) introduced earlier in this chapter. A baseball, flying through the air, has a propensity to shatter a pane of glass of a given kind if it is placed in the path of the ball. As the ball travels, however, it loses energy, and its propensity to shatter glass changes along its path. A photon of a given frequency, in contrast, retains the same propensity to interact with a hydrogen atom as long as it exists.

The next three examples, (12)–(14), illustrate y-type interactions. As I remarked in the preceding chapter, some readers may find it strange to regard these as causal interactions at all. I do not think it is illegitimate. However, whichever attitude one takes toward such cases, we can say that as a given process transpires, there is a probability that some change will occur, and that two processes will emerge. In the case of the carbon 14 atom, the situation is quantitatively well established. The half-life of carbon 14 is 5,730 years; this means that there is a probability of 1/2 that a y-type interaction of the sort mentioned in example (14) will occur in that period of time. The carbon 14 atom has a propensity of 1/2 to eject an electron in any period of 5,730 years. This is another case in which the propensity remains constant; atoms do not 'age.' Even if one insists upon saying that this is a spontaneous occurrence, rather than a causal interaction, it is equally correct to state that the causal process consisting of this atom transmits a propensity of the sort described. Hence processes of some kinds carry propensities to undergo changes even if no previously existing external process precipitates the change.

Interactions of the λ-type do not pose any additional problems, nor do they furnish much additional insight. If a photon of relatively low energy impinges upon an atom, it is absorbed, and the energy of the atom is changed from the ground state to a higher level. If a somewhat more energetic photon impinges upon a similar atom, the photon will be absorbed but the atom will be ionized. This constitutes an x-type interaction, for initially we have a neutral atom and a photon, while subsequently we have a free electron and a charged ion. If the photon is still more energetic, Compton scattering may take place.

The so-called propensity interpretation of probability has gained a good deal of popularity in recent years, and I am inclined to think that this idea of propensity, as a probabilistic disposition, is valuable. It is just such dispositions that seem to me to lie at the foundation of probabilistic causality.[6] There is, however, a strong reason for rejecting the notion that propensities constitute an adequate interpretation of the probability calculus, and consequently, for refusing to speak of a propensity interpretation of probability. The reason, basically, is that only some, but not all, of the probabilities with which we deal when we apply the probability calculus can reasonably be construed as propensities. The difficulty, which was

[6] (Fetzer, 1981, chaps. 4–6) offers an interesting and well-developed theory of causal explanation based upon propensities and probabilistic causes.

pointed out to me by Paul Humphreys, arises most clearly in connection with Bayes's theorem.

Consider, for example, a factory that produces can openers. There are only two machines, which we may designate A and B, in this factory. Machine A is ancient; it produces one thousand can openers per day, and 2.5% of these are defective. Machine B is more modern; it produces ten thousand can openers per day, and only 1% of its products are defective. Suppose, at the end of the day, that all of the defective can openers (which have been sorted out by the inspectors) are placed in a box. Someone randomly picks a can opener out of the box, and asks for the probability that it was produced by the modern machine B. We can easily calculate the answer; it is 4/5. Nevertheless, I find it quite unacceptable to say that this defective can opener has a propensity of 0.8 to have been produced by machine B. It makes good sense to say that machine B has a propensity of 0.01 to produce defective can openers, but not to say of the can opener that it has a certain propensity to have been produced by that machine. Propensities are causal probabilities, and, as such, they play an indispensable role in the probabilistic causal mechanisms of the universe. However, not all probabilities are causal in that way, and so it is a mistake, in my opinion, to try to *define* "probability" in terms of propensities (see Salmon, 1979, pp. 213–214).

8 | Theoretical Explanation

ONE OBVIOUS FACT about scientific explanations—and common-sense explanations as well—is that they frequently appeal to unobserved or unobservable objects. We explain diseases in terms of microorganisms. We explain the fertilization of plants on the basis of tiny particles of pollen transported by bees that we did not happen to observe. We explain television transmission by appealing to electromagnetic waves which propagate through space. We invoke DNA molecules to explain genetic phenomena. We explain human behavior in terms of neurophysiological processes or, sometimes, in terms of unconscious motives. In all of these examples, entities, processes, and events that are not observed—and that in many cases are not directly observable by the unaided human senses—are invoked for the purpose of explaining some phenomenon. The task of this chapter will be to discuss the status and role of objects of these kinds in scientific explanation.

From the outset it should be recalled—as announced in chapter 1—that I have no intention of raising the phenomenalism/realism issue. My point of departure for this discussion is physicalistic, and I shall assume without further ado that we can observe many macroscopic physical objects, processes, and events, and that we can legitimately infer the existence of such entities when they are not actually being observed. We infer the continuous existence of the kitchen clock while we are not at home, and explain the positions of its hands in terms of continuous processes that have transpired in our absence. We maintain that the planet Mars exists and moves in a continuous path during the day and at times when the sky is obscured by clouds. Things of these sorts are observable by an observer who is suitably placed, and we are confident about the outcomes of such observations were they to be made. If any serious question arises, it can in principle always be settled by making an appropriate kind of observation. This approach, which conforms to common sense, enables us to endow our world with a great deal of spatiotemporal continuity. In the light of what has been said in earlier chapters about causal processes, this aspect of the situation has considerable importance for our understanding of scientific explanation.

CAUSAL CONNECTIONS AND COMMON CAUSES

Let us begin by reconsidering one of our simple examples. Two students, Adams and Baker, submit essentially identical term papers in a particular course. There is, of course, the logical possibility that the two papers were produced entirely independently, and that the resemblance between them is a matter of pure chance. Given the overwhelming improbability of this sort of coincidence, no one takes this suggestion seriously. Three reasonable explanatory hypotheses are available: (1) Baker copied from Adams, (2) Adams copied from Baker, or (3) both copied from a common source. We can understand these possibilities in terms of the processes and forks discussed in chapters 5 and 6. There is either (1) a causal process running from Adams's production of the paper to Baker's, (2) a causal process running from Baker's production of the paper to Adams's, or (3) a common cause—for example, a paper in a fraternity file to which both Adams and Baker had access. In the case of this third alternative, there are two distinct causal processes running from the paper in the file to each of the two papers submitted by Adams and Baker, respectively.

As noted previously, there is a further alternative that we do not consider acceptable. Suppose that Adams and Baker are stars on the college basketball team, and that the team will win the championship in its conference if and only if at least one of these two players participates in the final game. Inasmuch as their plagiarism was discovered, however, both are disqualified and the team does not win the championship. This failure is a common effect of the two causes, Adams's disqualification and Baker's disqualification, but the common effect is not invoked to explain the resemblance between their papers. We reject the notion that such coincidences occur because nature conspires to bring about the defeat, or because events are teleologically 'drawn toward' future ends.

According to the viewpoint I am attempting to develop, statistical relevance relations require causal explanations. In order to explain these relations, let us introduce the concept of *causal relevance*. We shall need two kinds of causal relevance—direct and indirect. Let us say that we have *direct causal relevance* of one event to another if there is a causal process connecting them, and if that causal process is responsible for the transmission of causal influence from one to the other. Let us say that we have *indirect causal relevance* if the two events are results of a common cause as characterized in terms either of an interactive fork or a conjunctive fork. As I emphasized in the discussion of causal forks, a common cause—of either the conjunctive or the interactive variety—requires causal processes connecting the common cause to each of the separate effects. We can now say, quite generally, that statistical relevance relations are to be explained in terms of causal relevance relations.[1]

In the example of the two plagiarists, let us assume that careful investigation reveals that neither copied directly from the other. In that case, we hypothesize the existence of a common cause, and this hypothesis is amenable to direct observational confirmation. Moreover, the causal processes connecting the common cause to the effects are themselves observable; indeed, they were observed by the plagiarists. In many other circumstances, however, we postulate the existence of causal processes and/or common causes when they are not directly observable.

Let us first consider the inference to spatiotemporally continuous causal processes; it can be illustrated by simple examples. The philosopher C. J. Ducasse had a deep love of animals, and he would not have caused any of them unnecessary harm. At one time, he was bothered by a mouse in his basement. Rather than buying an ordinary mousetrap, which would have killed or injured the mouse, he carefully constructed a trap that would capture the mouse without hurting it. He set the trap and captured a mouse. On the way to his office the next morning, he released it in a vacant field some distance from his house. That night, he set the trap once more, and again he captured a mouse. He released it in the same vacant field. After this pattern of events had been repeated on a number of successive days, Ducasse had the feeling that the mouse he caught looked rather familiar. Before releasing it in the vacant field, he placed a small dab of white paint on its head. That night, he set the trap, and the following morning he found a mouse with a white mark on its head. That morning, he took the mouse a good deal farther before releasing it, and thereafter—for a while, at any rate—he had no mouse in his basement. Although the physical process, which consisted of the life of the mouse, was observed by Ducasse (or any other human we may suppose) only at disconnected times, we have no doubt that the process itself possessed spatiotemporal continuity. Ducasse employed the mark method to ascertain whether he was dealing with a single causal process (rather than many different mice), and there is no question about the spatiotemporal continuity of the transmission of the mark. This example is unproblematic, for the process is one which is in principle observable throughout its duration.

We go much farther in imputing spatiotemporal continuity to our world. For example, several years ago I had intermittent trouble with the reception on my television receiver. At quite irregular times, it seemed, the picture broke up into a herringbone pattern. Eventually, I found that the breakup

[1] Until we get to the quantum domain, at any rate. What happens there will be discussed in chapter 9.

of the picture was highly correlated with broadcasts made from a nearby police radio station. This correlation was explained by electromagnetic waves that traveled in a continuous path from the police transmitter to my television receiver. Electromagnetic radiation of frequencies employed in radio and television transmission are, of course, altogether imperceptible by human sense organs. Nevertheless, we feel entitled to appeal to such processes to provide spatiotemporally continuous causal connections among correlated events.

This example has considerable importance for our discussion because it raises the issue of action-at-a-distance in an appropriate scientific context. The electromagnetic theory developed by Faraday and Maxwell postulated the existence of electric and magnetic fields as well as the existence of electromagnetic radiation. After Hertz had experimentally confirmed the predictions of Maxwell's theory concerning the propagation of electromagnetic waves, there were, it seems to me, three main reasons for maintaining that a spatiotemporally continuous causal process connected the electric discharge at Hertz's source with the spark he observed in his detector. First, although the electromagnetic disturbance is not directly observable, if detecting devices were set up at intermediate points between his source and his original detector, they would register the presence of the disturbance at the intermediate points. No such detection would occur, of course, unless the source was creating a disturbance. Second, it takes a nonvanishing time interval for the disturbance (or any other change in the electric or magnetic field) to produce any effect at a distant point. This fact strongly suggests propagation from one place to the other in a continuous fashion at a finite velocity. Third, assuming the existence of fields, the behavior of a charged particle depends solely upon the nature of the field where the particle is located, as shown by the equation for the Lorentz force:

$$\mathbf{F} = q\mathbf{E} + \frac{q}{c}\mathbf{v} \times \mathbf{B}$$

where $\mathbf{E}$ and $\mathbf{B}$ are the electric and magnetic field vectors, respectively, at the location of the particle (which carries a charge q and moves with velocity $\mathbf{v}$). It does not depend in any way upon the arrangement of the distant sources which produce the field at that place.

Prior to the twentieth century, there seemed to be one compelling example of action-at-a-distance, namely, Newtonian gravitation. Some scientists speculated about the existence of a gravitational ether—analogous to the luminiferous ether—but with the demise of the luminiferous ether at the beginning of the present century, this suggestion lost most of its appeal. Contemporary gravitational theories do, however, hypothesize the

existence of gravitational fields, and gravitational radiation, which is propagated at a finite velocity. Although few, if any, physicists are confident
that gravitational waves have actually been detected as yet, there is indirect
evidence for the existence of gravitational radiation.[2] As long as we steer
clear of quantum mechanical phenomena, our current world picture seems
to incorporate the notion that causal influence is transmitted from one part
of space-time to another by means of spatiotemporally continuous processes. The specter of action-at-a-distance in quantum mechanics is a disturbing prospect; I shall comment upon it at the conclusion of this chapter.
Outside of the quantum domain, however, action-by-contact seems to be
the rule, and scientific theories that introduce unobservable processes to
conform to that pattern have met with considerable predictive and explanatory success.

Let us now turn to the postulation of common causes. In chapter 6 I
mentioned the 'twin quasars' 0975 + 561 A, B as a striking example. As
I said, the observation of the galaxy that acts as a gravitational lens—
actually, it would be more accurate to say that a cluster of galaxies is
involved—was rather fortuitous. Astronomers had, of course, observed
the quasar, which may be regarded as the common cause, but they could
not observe directly *that* they were seeing a single quasar. Moreover, the
quasar is only part of the cause, for the galaxy (or cluster) that produces
the gravitational bending is an essential part of the common cause. Even
before it had been observed, the need for some sort of common cause of
the spectacular correlation between the spectra of the two images was
acknowledged on all sides. The only doubt concerned the nature of the
common cause. Observation of the intervening elliptical galaxy settled that
question. In the few months between the writing of the initial draft of
chapter 6 and the writing of the initial draft of this chapter, another well-
documented instance of a gravitational lens has been found, and there is
now a strong presumption that many other cases of this phenomena are
observable—either with present earthbound telescopes or with the space
telescope, if it ever is put into orbit. This is an excellent example, I believe,
of the scientific power and success of the common cause principle.

Additional examples of scientific use of this principle abound. Coinci-
dence-counting techniques are a standard part of modern physics. In conducting the Compton scattering experiment, for example, one checks for
a correlation between photons scattered at a certain angle and electrons
ejected with a particular kinetic energy. The observed coincidences provide
an entirely satisfactory basis for inferring that *unobserved* collisions be-

<hr>

[2] For a recent nontechnical, yet authoritative, discussion of gravitational radiation, see
(Davies, 1980). A more technical treatment can be found in (Weinberg, 1972).

tween incident photons and (for all practical purposes) stationary electrons have occurred.

The appearance of two tracks in a cloud chamber emerging from a common point is taken as evidence of an event that constitutes a common cause. It may be, for example, a short-lived neutral K meson (which leaves no track) decaying into a positively charged pion and negatively charged pion.

If a crystal gazer in a neighboring town were able to predict with high reliability the outcome of the sixth race at the local track, we would be as confident of a common cause as we are in the case of the falling barometric reading and the storm. We might hunt in vain for the common cause, and for the processes that connect that cause with the outcome of the race and with the prediction made by the crystal gazer, but if the correlation were strong, we would not be shaken in our confidence that they exist. We might not be able to find out who fixes the race or how it is done. We might not be able to discover how the information is transmitted to the crystal gazer. Nevertheless, there would be no real doubt about the existence of the common cause and the causal processes.

EXPLANATORY VERSUS INFERENTIAL PRINCIPLES

As the various examples we have considered show, the principle of the common cause is used sometimes as an explanatory principle and sometimes as a principle of inference. In the case of the barometer, we are confident enough about the existence of the low atmospheric pressure to invoke the fact to *explain* the correlation between the falling barometric reading and the occurrence of a storm. In the case of the crystal gazer and the outcomes of horse races, we use the principle to *infer* the existence of the common cause, which we have not yet succeeded in locating. In the twin quasar example, the principle is used in both of these ways. Initially, recognition of the striking correlation between the spectra of the two images led to the *inference* that there must be some sort of common cause; later, the elliptical galaxy that was observed to be present in an appropriate place was invoked as a crucial part of the *explanation* of that correlation.

This dual use of the principle of the common cause is not at all unusual. Indeed, it seems to me, the fact that we are so often successful in employing the principle to infer the existence of a common cause that is subsequently discovered, and which then provides an adequate explanation of the apparent coincidence we wish to explain, rightly gives us enormous confidence in this important principle. Moreover, a more general statement seems warranted. We seem justified in claiming impressive success for the principle—stated earlier in this chapter—connecting statistical relevance to causal relevance: relations of statistical relevance between noncontiguous events are to be explained in terms of relations of direct or indirect causal relevance. Direct causal relevance involves the existence of spatiotemporally continuous causal processes; indirect causal relevance requires a common cause in addition to the connecting causal processes.

Let me illustrate by means of still another example that has some scientific importance. During the last few years, a malady popularly known as "Legionnaires' disease," which has broken out at several different times and places in the United States, has attracted a good deal of attention. Not long after the disease was first recognized in Philadelphia in 1976, *Legionella*, the bacterium responsible for the disease, was identified. One outbreak occurred in the Baptist Memorial Hospital in Memphis in 1978. Careful investigation revealed that a cooling tower of the hospital's air conditioning system harbored a colony of the bacteria; this was the common source of bacteria that infected thirty-nine people, seven of whom died as a result. It was further discovered that the bacteria were carried aloft in droplets of water from the cooling tower, and that these droplets entered the air intake of the hospital's cooling system, from which they were circulated into the hospital rooms. The bacteria, traveling spatiotemporally continuous trajectories, constitute the continuous causal processes linking the common cause with several correlated effects. (See *Science* 210 [November 14, 1980]:745–749.) I am inclined to think that the explanation would have been considered manifestly unsatisfactory if either the common cause or the continuous connecting causal processes had been lacking.

It is to be noted, of course, that neither the elliptical galaxy in the twin quasar example nor the bacteria in the example of Legionnaires' disease are visible to the naked eye; telescopes and microscopes, respectively, are needed to make the observations. This aspect of these examples is unproblematic. We have well-established optical theories to account for the operation of such devices, and these theories of geometric optics are adequately verified on the basis of macroscopic investigations. However, since the lower limit on the wavelength of visible light is about 4,000 angstroms, or 4×10^{-5} cm, while the diameter of an atom is of the order of 10^{-8} cm, there is no possibility, even in principle, of viewing atoms or simple molecules by means of an ordinary microscope. It is, nevertheless, common practice nowadays to invoke atoms and molecules, to say nothing of subatomic particles, to explain various phenomena. What justification have we for appealing to unobservable entities of these sorts?

Many contemporary philosophers would challenge the claim that these objects are unobservable. It is arbitrary, they might say, to place the limit of observability at the capability of an ordinary microscope using light to which humans are sensitive. Much smaller entities can be observed by means of such devices as scanning electron microscopes and bubble chambers; indeed, such observations of submicroscopic entities are made routinely in modern science. Now, I am by no means unsympathetic to this way of speaking, but I think we should give some consideration to the manner in which it can be justified. Therefore, for purposes of the present discussion, let us adopt a narrower definition of observability that will place atoms and simple molecules unequivocally into the category of unobservables. Taking this as our point of departure, let us see whether we can provide a satisfactory rationale for the use of such unobservables for explanatory purposes. The principle of the common cause will play a key role in this investigation.

THE COMMON CAUSE PRINCIPLE AND MOLECULAR REALITY

One great focus of interest in twentieth-century philosophy of science has been the controversy over instrumentalism and scientific realism. Major contributions to the discussion have been made by Mach, Russell, Duhem, Carnap, Reichenbach, Popper, Nagel, and Hempel—to name only a few. Moreover, the climate of opinion on this issue has changed radically. During the earlier decades of this century, various forms of instrumentalism seemed in general to hold sway; when Reichenbach and Popper defended realism in the thirties, they were departing drastically from the mainstream. In recent years, various forms of scientific realism have gained ascendancy, and instrumentalism has few important defenders. Although I had not joined the debate in print before 1978, I felt the tension between these two viewpoints; I felt my own sentiments moving away from instrumentalism toward scientific realism. At the same time, I have wondered whether the general shift in opinion has been more than merely a change in philosophical fashion, for the arguments that were offered in support of scientific realism have struck me as quite unconvincing. Bas van Fraassen has been the sharpest recent critic of the realist position, and I agree almost entirely with his criticisms of the prorealist arguments. I began to suspect that what moved me (and many other philosophers as well, perhaps) was not philosophical argument, but, rather—to use Russell's apt phrase—growth of a more "robust sense of reality." It is, of course, disquieting to find that one's main philosophical commitments arise—not from reasoned argument—but from purely psychological considerations.

In an effort to alleviate this intellectual discomfort, I decided to try an empirical approach to the philosophical problem. Since it seemed unlikely that scientists would have been moved by the kinds of arguments supplied by philosophers, I felt that some insight might be gained if we were to consider the evidence and arguments that convinced scientists of the reality of unobservable entities. Although scientists, by and large, seem committed to the existence of a variety of unobservable entities—for example, the planetary orbits of Copernicus and Maxwell's electromagnetic waves (outside of the range of visible light)—the existence of atoms and molecules, as the microphysical constituents of ordinary matter, is the most clear and compelling example. It is fortunate that a superb account of this historical case is provided by Mary Jo Nye in *Molecular Reality* (1972). It is also fortunate that the French physical chemist Jean Perrin, who played a central role in this historical development, provided a lucid account in his semi-popular book, *Les Atomes* (1913). Examination of these works reveals a clear-cut form of argument.

As everyone knows, a primitive form of atomism was propounded in antiquity by Democritus, Lucretius, and others, but no very strong empirical evidence in support of such a theory was available before the beginning of the nineteenth century. However, from the time of Dalton's work (early in that century) until the end of the century, the atomic hypothesis was the subject of considerable discussion and debate. During this period, responsible, well-informed scientists could reasonably adopt divergent viewpoints on the question. Michael Gardner (1979) has given an enlightening discussion of these nineteenth-century developments, relating them to the realism/instrumentalism issue. Within a dozen years after the turn of the century, however, the issue was scientifically settled in favor of the atomic/molecular hypothesis, and the scientific community—with the notable exception of Ernst Mach—appears to have been in agreement about it.

The decisive achievement was the determination of Avogadro's number N, the number of molecules in a mole of any substance. This number is *the* link between the macrocosm and the microcosm; once it is known a variety of microquantities are directly ascertainable. Loschmidt gave what seems to be the first crude estimate of N in 1865; at the turn of the century, rough approximations based upon the kinetic theory of gases were available. Perrin's chief experimental efforts were devoted to the precise ascertainment of N through the study of Brownian movement—the random agitation of microscopic particles (e.g., pollen or smoke) suspended in fluid. The difficulties that stood in the way of success in such a venture were awesome, and Perrin's achievement is a milestone in the emergence of modern microphysics. Since Nye's excellent account is available, I shall not attempt to provide a historical treatment of this topic, but I shall give a brief chronology of some major developments in the first few years of the twentieth century.

Perrin, who was born in 1870, recognized early in his career the crucial importance of providing a scientifically convincing demonstration of the reality of atoms and molecules. The achievement of this aim constituted the major motivation for his research, at least until 1913. His doctoral dissertation was on X rays and cathode rays—topics that were at the forefront of microphysics just before the turn of the century. Shortly after receiving his doctorate in 1897, he was put in charge of a course in physical chemistry at the Sorbonne. In organizing the materials for this course, he became interested in various aspects of thermodynamics and the micro-structure of matter that eventually led to his work on Brownian movement. In 1901, in commenting upon the then-current debate between the 'partisans of the plenum' and the 'partisans of the void' (atomists), he explicitly noted the common view that the controversy was a matter of philosophical speculation that could never be settled by experimental evidence. At the same time, he pointed to the possibility of verifying physical statements about objects whose dimensions fall far below 0.25 microns, the limit of resolution of an ordinary microscope (Nye, 1972, p. 83).

A good deal of Perrin's early research concerned the nature of colloids—suspensions of particles larger than molecules but still too small to be discerned under an ordinary microscope. In 1903, Siedentopf and Zisgmondy published an account of the ultramicroscope, which made possible the observation of particles with diameters as small as 5×10^{-3} microns. With this amount of magnification, the Brownian movement of colloidal particles could be viewed. Perrin's earliest determination of a precise value for Avogadro's number was based upon observations of the vertical distribution of particles in colloidal suspensions.

In 1905–1906, Einstein and Smoluchowski provided the first adequate theoretical accounts of the phenomenon that Robert Brown had observed early in the nineteenth century. At about the same time (1906), Perrin maintained that particles undergoing Brownian motion reveal *qualitatively* the nature of the fluid in which they are suspended (Nye, p. 85). The Einstein-Smoluchowski theory had little, if any, influence upon the experimental work of Perrin that culminated in his first precise determination of Avogadro's number (Nye, p. 111), but by the time these results were published (1908), Perrin realized that they constituted *quantitative* experimental confirmation for the Einstein-Smoluchowski theory. Einstein was pleased, but also astonished, by Perrin's achievement, for he had held the opinion that the practical difficulties standing in the way of such experiments would be insuperable (Nye, p. 135).

Perrin's experimental achievement was, indeed, monumental. The experiments required the preparation of tiny spheres—less than 1 micron in diameter—of gamboge, a bright yellow resin. To serve adequately in the experiments, these spheres had to be uniform in size and density, and the size had to be measured with great precision. Vast numbers of painstaking observations were needed. The results were presented in four papers that were published in the *Comptes rendus* of the Académie des Sciences in 1908. It is of the greatest importance to *our* story to note that these papers included not only the precise value of Avogadro's number ascertained on the basis of his study of Brownian movement, but also a comparison of that value with the results of several other determinations based upon entirely different methods, including Rutherford's study of radioactivity and Planck's work on blackbody radiation (Nye, pp. 109–111).

Perrin's 1908 results had an immediate impact in some quarters. Ostwald, who was one of the most prominent and staunch opponents of the atomic/molecular hypothesis, had stated categorically in 1906 that "atoms are only hypothetical things" (quoted by Nye, p. 151). However, in the fourth edition of his *Grundiss der physikalischen Chemie* (1908), Ostwald did an about-face, writing:

> I have satisfied myself that we arrived a short time ago at the possession of experimental proof for the discrete or particulate nature of matter— proof which the atomic hypothesis has vainly sought for a hundred years, even a thousand years. . . . [The results] entitle even the cautious scientist to speak of an experimental proof for the atomistic constitution of space-filled matter. (Quoted by Nye, ibid.)

Ostwald is referring here to the work of J. J. Thomson on the kinetic theory of gases, as well as to Perrin's work on Brownian movement.

Perrin's work, up to this time, had dealt with translational Brownian movement. Recognizing that the principle of equipartition implies that thermal energy also produces rotations, he made further determinations of N on the basis of the rotational motions of Brownian particles. In order to carry out this experiment, he had to create imperfect spheres of gamboge, for the rotational motion of a perfect sphere is impossible to observe. With characteristic ingenuity, he accomplished this feat. In a 1909 paper, Perrin reported not only his own ascertainments of Avogadro's number by three distinct methods—namely, the vertical distribution of Brownian particles, the translational motion of Brownian particles, and the rotational motion of Brownian particles—but also eight other distinct determinations of N by other investigators (Nye, pp. 133–135). The results were all in striking agreement with one another.

In 1913, Perrin published *Les Atomes*, in which he recounted his experimental efforts and summarized the evidence for the reality of molecules. Instead of focusing upon one or two beautiful experimental determinations of precise values of Avogadro's number, he lays great stress upon the fact that N has been ascertained in a considerable number of independent ways—in fact, he lists thirteen distinct methods (Perrin, 1913,

p. 206; Nye, p. 161). In her careful historical analysis of the situation, Nye also places great emphasis upon the variety of independent methods of ascertaining N (pp. 110–111, 133–135). At a 1912 conference, Poincaré likewise emphasized the variety of independent determinations (Nye, p. 157). Since I shall be arguing that this variety plays a decisive role in the argument, let me give a somewhat oversimplified indication of its range. Emulating the venerable tradition established by St. Thomas, I shall offer five ways.

1. *Brownian movement.* If one applies the principle of equipartition to the Brownian particles, then it follows immediately that the average kinetic energy of the Brownian particles is equal to the average kinetic energy of the molecules of the fluid in which they are suspended. This is a straightforward interpretation of the claim that the system is in thermal equilibrium. Molecular velocities are, in principle, directly measurable; moreover, they can be derived from macroscopic quantities via Herapath's formula

$$P = \tfrac{1}{3}\rho v^2$$

where P is the pressure and ρ the density of a sample of gas. In addition, it is possible in principle to measure both the mass and the velocities of Brownian particles. From these quantities it is easy to calculate the mass of a molecule of the fluid, and hence, Avogadro's number, which is the molecular weight divided by the mass of a single molecule.

In fact, this simple approach is impractical; because the Brownian particle changes its direction of motion with extremely great frequency, its velocity cannot be measured directly. Perrin found it necessary to take a less direct approach. In his first experiments, he examined the vertical distribution of Brownian particles in suspension under equilibrium conditions. Such particles are subject to two opposing forces. Given that the particles are denser than water, the gravitational force makes them tend to sink to the bottom. A force due to osmotic pressure (a diffusion process) tends to raise them. By carefully ascertaining the exponential distribution according to height, Perrin was able to calculate Avogadro's number. This result is, in principle, similar to the simpler idea outlined in the preceding paragraph, for it leads to values for the average kinetic energy of the Brownian particles and of the molecules of the fluid. (For details see Nye, 1972, pp. 102–111).

2. *Alpha decay.* During the first decade of the present century, it was discovered that alpha particles are helium nuclei. Alpha particles were detected by scintillation techniques; it was, consequently, straightforward to ascertain the rate at which a given substance—radium, for example— produces helium nuclei. Rutherford recognized that the alpha particles

quickly pick up electrons, transforming them into helium atoms, and that the helium generated by a given quantity of radium in a fixed period of time could be captured and weighed. In this way, it was possible to count the number of atoms required to make up a given mass of helium. Avogadro's number comes directly from such data.

3. *X-ray diffraction*. M. von Laue had the idea that the arrangement of atoms in a crystal could serve as a diffraction grating for X rays. Given a knowledge of the wavelengths of the X rays, it is possible to examine the diffraction patterns and calculate the spacing between the atoms. With this information, one can ascertain the number of atoms in a given crystal. Experiments of this sort were actually conducted in 1912 by W. Friedrich and P. Knipping. Avogadro's number follows directly.

4. *Blackbody radiation*. In the closing days of the nineteenth century (December 14, 1900), Max Planck presented his derivation of the law of blackbody radiation, from which he derived the relation

$$\lambda_{max} \times kT/c = 0.2014 \times h$$

where k is Boltzmann's constant, T the temperature of the blackbody, λ the wavelength at which the maximum energy is being radiated, c the speed of light, and h Planck's constant. From this relation, if h is known, we can derive the value of Boltzmann's constant, since all of the other quantities can be measured macroscopically. The speed of light was well known by the end of the nineteenth century.

There are various ways of ascertaining a value for Planck's constant; one obvious method comes from Einstein's theory of the photoelectric effect. If we let E stand for the total energy that a photon can transfer to an electron—that is, the maximum kinetic energy of photoelectrons plus the work function of the metal upon which the photons impinge—we have h as a constant of proportionality between E and v, the frequency of the incident light.

During the nineteenth century, the ideal gas law

$$PV = nRT$$

was derived from the kinetic theory of gases, and the universal gas constant R was measured empirically. Since n represents the number of moles of the ideal gas, R is the gas constant *per mole*. Boltzmann's constant k is the gas constant *per molecule*. Hence $R/k = N$.

5. *Electrochemistry*. The faraday F is the amount of charge (96,500 coulombs) required to deposit by electrolysis one mole of a monovalent metal—for example, silver—on an electrode. It takes the charge of one electron to deposit each ion, that is,

$F/e = N.$

The experimental determination of e, the charge of the electron, by Robert A. Millikan in 1911, thus furnished another way of ascertaining Avogadro's number.

The foregoing list of ways to establish the value of Avogadro's number does not coincide with Perrin's list, but it gives a fair indication of the variety of methods available in the first few years of the present century.[3] As I have already remarked, the variety of these approaches is remarkable and striking, and it was fully appreciated at the time. At the conclusion of a 1912 conference at which the atomic/molecular hypothesis was a main topic of discussion, Poincaré remarked:

> A first reflection is sure to strike all the listeners; the long-standing mechanistic and atomistic hypotheses have recently taken on enough consistency to cease almost appearing to us as hypotheses; atoms are no longer a useful fiction; things seem to us in favour of saying that we see them since we know how to count them. . . . The brilliant determinations of the number of atoms by M. Perrin have completed this triumph of atomism. . . . In the procedures deriving from Brownian movement or in those where the law of radiation is invoked . . . in . . . the blue of the sky . . . when radium is dealt with. . . . The atom of the chemist is now a reality. (Quoted by Nye, 1972, p. 157)

In *Les Atomes*, published in the next year, Perrin also places considerable weight upon the variety of methods. Immediately following his table of thirteen distinct experimental approaches to Avogadro's number, he writes:

> Our wonder is aroused at the very remarkable agreement found between values derived from the consideration of such widely different phenomena. Seeing that not only is the same magnitude obtained by each method when the conditions under which it is applied are varied as much as possible, but that the numbers thus established also agree among themselves, without discrepancy, for all the methods employed, the real existence of the molecule is given a probability bordering on certainty. (Perrin, 1913, pp. 215–216)

It seems clear that Perrin would not have been satisfied with the determination of Avogadro's number by any single method, no matter how carefully applied and no matter how precise the results. It is the "re-

---

[3] An excellent discussion of the significance of Avogadro's number N, and the way in which it serves to link the macrocosm and the microcosm, can be found in (Wichmann, 1967, chap. 1). Further details regarding several of the previously mentioned methods used to ascertain the value N are presented there.

markable agreement'' among the results of many diverse methods that supports his conclusion about the reality of molecules.

If there were no such micro-entities as atoms, molecules, and ions, then these different experiments designed to ascertain Avogadro's number would be genuinely independent experiments, and the striking numerical agreement in their results would constitute an utterly astonishing coincidence. To those who were in doubt about the existence of such micro-entities, the ''remarkable agreement'' constitutes strong *evidence* that these experiments are not fully independent—that they reveal the existence of such entities. To those of us who believe in the existence of these sorts of micro-entities, their very existence and characteristics—as specified by various theories—*explain* this ''remarkable agreement.''

The claim I should like to make about the argument hinted by Poincaré, and stated somewhat more explicitly by Perrin, is that it relies upon the principle of the common cause—indeed, that it appeals to a conjunctive fork. We can say, very schematically, that the coincidence to be explained is the ''remarkable agreement'' among the values of N that result from independent determinations. The situation is, I think, quite analogous to the testimony of independent witnesses. Suppose that a murder has been committed, and that one crucial factor in the investigation concerns whether a particular suspect, John Doe, was in the vicinity of the crime at about the time it was committed. Suppose that several different witnesses testify, not only to his whereabouts, but also to a variety of details—such as how he was dressed, what he was carrying, and the fact that he arrived in a taxi. Indeed, suppose that all of these witnesses were able to give the license number of the taxi. If collusion can be ruled out, and if we can be confident that they were not coached, then the agreement of the witnesses on details of the foregoing sorts would constitute strong evidence that they were reporting facts that they had observed. Moreover, even if none of the witnesses could be considered particularly reliable, the agreement among their reports would greatly enhance our confidence in the veracity of their account. It would be too improbable a coincidence for all of them to have fabricated their stories independently, and for these stories to exhibit such strong agreement in precise detail.

For the sake of argument, let us suppose that there are five witnesses, each of whom claims to have observed the arrival of John Doe at the scene of the crime. Each claims to remember the license. Assume that any license consists of two letters and four digits. Since there are over six million different combinations, the probability that five independent witnesses would choose the same combination by chance is too small to consider seriously. We might think about the question of filling the blank in the following statement: The number of the license on the taxi is ————. If each of the witnesses fills the blank in the same way, we do not believe it can be due to mere chance coincidence. If the witnesses had not been present to see the suspect arrive in the taxi, or if they had been present but no taxi had been involved, we could hardly expect such agreement in testimony.

The situation pertaining to Avogadro's number can be viewed in much the same way. Suppose we have five scientists. One of them is studying Brownian movement (1), another is studying alpha decay (2), another is doing X-ray diffraction experiments on crystals (3), still another is working on the spectrum of blackbody radiation (4), while the remaining one is doing an experiment in electrolysis (5). Notice what a wide variety of substances are involved and how diverse are the phenomena being observed: (1) random movements of tiny spheres of gamboge suspended in water are viewed under a powerful microscope; (2) scintillations on a screen exposed to alpha radiation are carefully counted, and the quantities of helium generated in the presence of a radioactive substance are carefully measured; (3) spatial relations among the light and dark areas on a photographic film that has been exposed to X-radiation are measured; (4) the light from the mouth of a blast furnace is spectroscopically separated on the basis of wavelength for the measurement of intensity; and (5) metallic silver is observed to collect upon an electrode placed in an electrolytic solution. These experiments seem on the surface to have nothing to do with one another. Nevertheless, we ask each scientist to fill in the blank in this statement: On the basis of my experiments, assuming matter to be composed of molecules, I calculate the number of molecules in a mole of any given substance to be —————. When we find that all of them write numbers that, within the accuracy of their experiments, agree with 6×10^{23}, we are as impressed by the "remarkable agreement" as were Perrin and Poincaré. Certainly, these five hypothetical scientists have been counting entities that are objectively real.

Let us consider the alternative hypothesis. At the turn of the century, the main hypothesis that was opposed to the atomic theory—one which rejected the kinetic-molecular theory of gases—was a phenomenological view. According to this theory, often known as energeticism, matter that appears under close scrutiny to be continuous is in fact continuous; it is not composed of myriad tiny corpuscles and (in the case of gases, at least) a great deal of empty space. Proponents of the phenomenological view maintained that the laws of phenomenological thermodynamics are strictly true down to the microscopic and submicroscopic levels. Ostwald, in particular, was greatly impressed by the fact that energy is a conserved quantity, and that the thermodynamic equations governing the behavior of energy are differential equations. He took these considerations to be a basis for doubting the hypothesis that matter is composed of discrete corpuscles, and as support for the view that energy, which is continuous, constitutes the underlying reality. Arguments of the foregoing sort were frequently coupled, among opponents of atomism, with extreme positivistic arguments about the unknowability of the fine structure of matter. Proponents of the kinetic-molecular theory, in contrast, were forced to regard at least some of the basic laws of thermodynamics—especially the second law—as statistical, and consequently, to admit that the phenomenological laws were subject to violation at the molecular and microscopic levels. The dispute about 'molecular reality' clearly involved a strange mixture of factual and philosophical components. One is reminded of the man, accused of drunken behavior, who protests, "Everyone knows I am a teetotaler, and besides, I only had two drinks."

Now suppose, for instance, that carbon is not composed of atoms. We take various forms of carbon—a piece of coal and diamond, for example—and try to count the constituent particles. We pulverize the coal and the diamond in whatever ways we can devise. It seems plausible to suppose that the results of such endeavors would *not* yield constant numbers for 12 grams of coal if we tried it many times, or for 12 grams of diamonds. It would be *most* unreasonable to expect numerical agreement between the results of counting coal particles and those of counting diamond fragments. Likewise, if water and alcohol were continuous fluids on every scale, no matter how fine, then we should not expect consistent results if we attempt to create fine sprays and count the number of droplets. These suggestions are, of course, egregiously crude from an experimental standpoint. Nevertheless, it seems to me, regardless of how carefully and ingeniously they were refined, we would not get the kinds of consistent numerical results that have actually emerged in the various ascertainments of Avogadro's number unless matter really does possess an atomic/molecular structure. If matter were ultimately continuous, the number of particles would depend solely upon the method by which the substance is broken into pieces, for the substance itself is not in any straightforward sense *composed of* constituent particles. What we find, in fact, is that when we try, in any of a variety of suitable ways, to count the number of atoms in 12 grams of carbon—whether it be in the form of coal, graphite, lampblack, or diamond—the number is always the same. Moreover, that number is the same as the number of water molecules in 18 grams of water, or the number of helium atoms in 4 grams of helium. Still other substances yield the same number of molecules per mole. These results apply to matter in the gaseous, liquid, and solid states, and with respect both to physical and to chemical properties. Such numerical consistency would be an unthinkably improbable coincidence if molecules, atoms, and ions did not exist.

I maintained previously that the argument by which the existence of atoms and molecules was established to the satisfaction of virtually all physical scientists in the first dozen years of the present century has the form of a common cause argument, and that it relies upon the structure of the conjunctive fork. We should recall explicitly at this point that conjunctive forks differ in a fundamental physical way from interactive forks. In the interactive fork, we have two or more causal processes that intersect spatiotemporally with one another. In the conjunctive fork, by contrast, we find causal processes that are physically independent of one another—that do not even intersect—but that arise out of common background conditions. The statistical dependency among the effects is a result of common background conditions. Remember, for instance, the victims of mushroom poisoning; their common illness arose from the fact that each of them consumed food from a common pot. Similarly, I think, the agreement in values arising from different ascertainments of Avogadro's number results from the fact that in each of the physical procedures mentioned, the experimenter was dealing with substances composed of atoms and molecules—in accordance with the theory of the constitution of matter that we have all come to accept. The historical argument that convinced scientists of the reality of atoms and molecules is, I believe, philosophically impeccable.

In presenting the idea of a conjunctive common cause, I emphasized the fact that there must be causal processes leading from the causal background to the correlated effects, and that a set of statistical relations—(1)–(4) of chapter 7—must be satisfied. There is little difficulty, I think, in seeing that the required causal processes do connect the existence and behavior of the micro-particles with the experimental results that furnish the basis for the calculation of Avogadro's number. In the case of X-ray diffraction, for example, electromagnetic radiation of known wavelength is emitted from an X-ray source, it travels a spatiotemporally continuous path from the source to the crystal, where it interacts with a grating in a way that is well established in optics. The diffracted radiation then travels from the crystal to a photographic plate, where another causal interaction occurs, yielding an interference pattern when the plate is developed. In the case of Brownian movement, the observed random motion of the Brownian particle is caused by a very large number of collisions (causal interactions) with the molecules of the fluid in which the particle is suspended. The other approaches to Avogadro's number patently involve similar sorts of causal processes and interactions. Indeed, it is by virtue of the causal properties of atoms, molecules, and ions that the various experimental ascertainments of N are in principle possible. The causal processes that lead from the microstructure of matter to the observed phenomena that yield values of N are quite complex, but our physical theories provide a straightforward account of them.[4]

Let us now try to see how the statistical relations used to define conjunctive forks apply to the ascertainment of N; for ease of reference, these relations are here repeated:

$$P(A.B|C) = P(A|C) \times P(B|C) \qquad (1)$$
$$P(A.B|\bar{C}) = P(A|\bar{C}) \times P(B|\bar{C}) \qquad (2)$$
$$P(A|C) > P(A|\bar{C}) \qquad (3)$$
$$P(B|C) > P(B|\bar{C}) \qquad (4)$$

In order to apply these formulas directly, we shall confine our consideration to two methods—alpha radiation and Brownian motion—though the formulas could obviously be generalized to apply to five or thirteen or any desired number of methods. We need to exercise a bit of care in assigning meanings to the variables A, B, and C. Let us suppose that A and B stand for classes of individual experiments involving alpha radiation and Brownian motion, respectively, that yield sufficiently accurate values for Avogadro's number. In Perrin's previously mentioned table of thirteen methods, the values of N range from 6×10^{23} to 7.5×10^{23}; the currently accepted value is 6.022×10^{23}. Let us say that values lying between 4×10^{23} and 8×10^{23} are acceptable. Given the difficulty of the experiments, we should by no means expect uniformly successful results. Next, let us understand C to represent experiments actually conducted under the initial conditions supposed to obtain in the case at hand. In the experiments on alpha radiation and helium production, for example, C would involve specification of the correct atomic weight for helium and the correct decay rate for the radioactive source of alpha particles. A mistake on either of these scores would provide us with an instance of $\bar{C}$. Similarly, if alpha particles did not come from the disintegration of atoms and if helium did not have an atomic structure, we would not have an instance of C. In the experiments on Brownian motion, if we suppose that the particles of gamboge are suspended in ordinary water, they must not be suspended in alcohol or heavy water. Likewise, we must have accurate values for the size and density of the Brownian particles. Given an accurate value for the molecular weight of the suspending fluid, and accurate values of the parameters characterizing the spheres of gamboge, we may say that C is satisfied. Of course, C would fail radically to be fulfilled if suspending fluids were not composed of molecules, but were strictly continuous media.

---

[4] It goes almost without saying, of course, that none of the methods used to ascertain the value of Avogadro's number does so exclusively on the basis of directly observable quantities without any appeal to auxiliary hypotheses. As Clark Glymour argues in detail in (Glymour, 1980, chap. 5), this fact does not undermine the validity of the argument.

With this general understanding of the interpretation of formulas (1)–(4), it is easy to see that these relations are satisfied. Equation (1) holds because the experiments on Brownian motion are physically separate and distinct from those on alpha radiation and helium production. The most famous ones were conducted in different countries. On the assumption that the initial conditions for these experiments have been satisfied, the probability of a successful outcome of an experiment of the one type is uninfluenced by successes or failures in experiments of the other type. Even if Rutherford and his associates in England had been inept experimentalists, that would not have had any influence upon the results of experiments conducted by Perrin in France, and conversely. Similarly, on the assumption that experiments of either type are conducted under incorrectly assigned initial conditions, it would seem reasonable to suppose that a successful outcome would simply be fortuitous, and that a lucky correct result on an experiment of one type would have no influence upon the chance of lucky correct results on experiments of the other. Thus equation (2) appears also to be satisfied.

There is, however, an important ground for uneasiness about relation (2). We are especially interested in one particular sort of mistake about initial conditions—namely, the mistake of supposing that material objects are composed of molecules and atoms when, in fact, they are not. Since material objects *are* composed of molecules, experiments cannot actually be conducted under these conditions. The reference class of such experiments is empty; hence the probabilities that occur in equation (2) *might seem* not to be well defined if we confine our attention to experimental conditions that fail to be instances of C for that reason. This consequence does not necessarily follow. If we can provide theoretical reasons for asserting what the outcome would be if the kinetic-molecular theory were mistaken, then, it seem to me, we can counterfactually assign values to the probabilities in (2).[5] Such assignments will clearly depend heavily upon the alternative hypotheses concerning the structure of matter that are available and regarded as serious contenders.

Consider Brownian motion. Perhaps water is a continuous medium that is subject to internal microscopic vibrations related to temperature. Suspended particles might behave in just the way Brown and others observed. Moreover, these vibrations might bear some relationship to the chemical properties of different substances that we (mistakenly) identify with molecular weight. Conceivably, such vibratory behavior might lead to precisely the results of experiments on Brownian motion that Perrin found,

[5] In (Reichenbach, 1954, Appendix) a way is proposed for assigning well-defined probability values in certain kinds of cases in which the reference classes are empty.

leading us (mistakenly) to think that we had successfully ascertained the number of molecules in a mole of the suspending fluid. It is for this reason that no *single* experimental method of determining Avogadro's number, no matter how ingeniously and beautifully executed, could serve to establish decisively the existence of molecules. The opponent could plausibly respond that, with respect to Brownian motion, fluids behave merely *as if* they were composed of molecules, and, indeed, as if they contained a specific number of molecules.

When we turn to helium production by alpha radiation, we must find another alternative explanation. Assuming that radium is *not* composed of atoms, we may still suppose that tiny globs of stuff are ejected from a mass of radium at a statistically stable rate. We may further claim that, for some reason, all of these globs are the same size. Moreover, we may assert that the globs are bits of helium that can be collected and weighed. We find, consequently, that it takes a given number of little globs of helium to make up 4 grams. Even given all of these assumptions, it is extremely implausible to suppose that the number of globs in 4 grams of helium should equal the number that Perrin (mistakenly) inferred to be the number of molecules in 18 grams of water. It is difficult—to say the least—to see any connection between the vibratory motions of water that account for Brownian motion on the noncorpuscular hypothesis and the size of the globs of helium that pop out of chunks of radium. And even if—in some fashion that is not clear to me—it were possible to provide a noncorpuscular account of the substances involved in experiments of these two types, we would still face the problem of extending this noncorpuscular theory to deal, not just qualitatively, but also quantitatively with the results of the many other types of experiments to which we have made reference. To scientists like Ostwald, the task must have seemed insuperable.

The claim that inequalities (3) and (4) are satisfied is quite straightforward. The chances of obtaining the appropriate results if the initial conditions are satisfied must surely be greater than the chances of obtaining them fortuitously under experimental conditions from which they should not have been expected to arise. It seems to me that, in a rough and schematic way at least, we have justified the assertion that the overall structure of the argument that was taken by most scientists early in the present century to establish conclusively the existence of 'molecular reality' is that of the conjunctive common cause.

Van Fraassen (1980, p. 123) has criticized my earlier appeal to the common cause principle for support of theoretical realism (in Salmon, 1978) on the ground that the effects that are correlated—the values of Avogadro's number derived from different experiments—cannot be traced back to any *event* that serves as a common cause. The fact that such results arise out of an experiment on Brownian motion and an experiment on electrolysis shows, at best, that there is some common feature of the conditions under which these experiments are performed. Quite so. In my previous exposition of this argument for theoretical realism, I was not clear on this crucial distinction between interactive forks and conjunctive forks; consequently, van Fraassen's criticism of that discussion is entirely well-founded. In chapter 6 of the present book, I tried to exhibit clearly this feature that all conjunctive forks (not including perfect forks) seem to share. I am extremely grateful to van Fraassen for forcing me to see that there are these two distinct kinds of causal forks, and for making me recognize that conjunctive forks involve physically distinct (nonintersecting) processes arising out of common background conditions. These considerations strengthen and clarify the use of common cause arguments on behalf of scientific realism.

THE EXPLANATORY POWER OF THEORIES

In the preceding discussion of the arguments supporting the existence of atoms and molecules, the common cause principle was used inferentially. Once we have satisfied ourselves that micro-entities of this sort exist, they can be invoked to explain many phenomena. In introductory physics textbooks and courses, the kinetic-molecular theory is used to *derive* the ideal gas law,

$$PV = nRT,$$

and that theory is also said to *explain* the ideal gas law. Indeed, this explanation is often presented as a prime example of theoretical explanations of empirical regularities. I am in strong agreement with this assessment, although, of course, I deny that the derivation *is* the explanation.

It will be recalled, I hope, that the ideal gas law was cited in an earlier chapter as an example of a noncausal regularity that has little, if any, explantory power of its own, but that cries out to be explained. The kinetic-molecular theory furnishes the required explanation. The ideal gas law provides a striking illustration of a statistical relevance relation—that is, the correlation among the values of pressure, volume, and temperature for a given quantity of gas—and the kinetic-molecular theory furnishes a *causal* explanation of that statistical regularity. The explanation is causal because a macroscopic sample of gas is taken to consist of an extremely large number of molecules, each of which qualifies as a causal process as we are construing that concept. Molecules can be marked; for example, radiation impinging upon a molecule can put it into an excited state, and it will remain in that state for some time after the interaction has occurred.

These causal processes—the molecules—enter into many causal interactions as they collide with one another and with the walls in their container. To most nineteenth-century kinetic theorists, the causal interactions were strictly deterministic, but we can cheerfully admit that they may actually be irreducibly statistical. Our causal concepts admit irreducibly statistical features without any strain. The kinetic-molecular theory also provides a statistical explanation for the second law of thermodynamics, for the phenomenon of Brownian motion, and—via Avogadro's law—for the empirical regularity that the distance required by an airplane for takeoff increases with increasing humidity of the atmosphere.[6]

The experimental proof—to use Ostwald's phrase—of the reality of atoms and molecules constitutes an epoch-making event in the history of scientific explanation. The explanatory successes that have followed upon the appeal to such micro-entities as atoms, molecules, ions, and electrons—in biology as well as physics and chemistry—provide a powerful body of inductive support, I believe, for the general conception of causal explanation I have been advocating in this book. We are, to a large degree, justified in the view that relations of statistical relevance are in principle amenable to causal explanation on the basis of relations of direct or indirect causal relevance. Direct causal relevance involves spatiotemporally continuous causal processes. These processes constitute the mechanism by which causal influence is propagated in our universe. Indirect causal relevance involves common causes of either the conjunctive or interactive variety. Events that are statistically correlated, but that are not joined by relations of direct causal relevance, are in principle amenable to explanation in terms of common causes. Scientific experience provides strong support for the appeal to unobservable common causes and causal processes when observable domains do not furnish the required causal connections. The common cause argument for the existence of molecules and atoms, which appeals to the multiplicity of apparently independent ascertainments of Avogadro's number, legitimizes the appeal to unobservables for the purpose of providing causal explanations of observable phenomena—especially explanations of empirical regularities.

There is, unfortunately, one major physical domain in which the foregoing approach does not seem to apply. As everyone knows, the years from 1900 to 1913, during which the kinetic-molecular theory was fully vindicated, also marked a period in which microphysics was beginning to undergo a profound revolution. During those years, Planck produced the quantum theory of blackbody radiation, Einstein published his quantum account of the photoelectric effect, and Bohr introduced his quantized

---

[6] The explanation of this regularity will be given in chapter 9, pp. 268–270.

model of the hydrogen atom. In the early 1920s with the work of Compton and de Broglie, the wave-particle duality for both light and matter was firmly established, and in the mid-1920s a full-fledged quantum mechanics emerged. This new theory was strange indeed; physicists and philosophers have been trying ever since to understand it. Bohr himself was one of the most influential contributors to this discussion of the foundations of quantum mechanics.

In an attempt to come to terms with the conceptual consequences of the new physical theory, Bohr enunciated his well-known principle of complementarity. It has been stated in a wide variety of ways, but one formulation has a direct bearing upon our discussion. According to this version of the principle, causal descriptions and space-time descriptions of quantum mechanical systems are complementary—hence it is not possible consistently to apply descriptions of both types to any given system at the same time. This means that microphysics is fraught with causal anomalies; causal explanations in terms of spatiotemporally continuous processes are precluded in principle. The quantum domain is not merely indeterministic; the account of probabilistic causality given in the preceding chapter can easily accommodate lack of deterministic causality. The problem is far more serious. It looks as if we may find correlations between noncontiguous occurrences that cannot be connected by continuous causal processes. The specter of action-at-a-distance arises from the microcosm. In the next chapter, I shall try to come to terms with this situation—at least to a very limited extent.

EMPIRICISM AND REALISM

The most important challenge to theoretical realism in recent years has come from van Fraassen, mainly in his provocative book *The Scientific Image* (1980), where he advocates a view called ''constructive empiricism.'' The key tenet of this position is that the acceptance of scientific theories involves the judgment that they are empirically adequate, but not necessarily that they are literally true when they seem to make reference to unobservable entities. He would maintain, for example, that the thermodynamics based upon the molecular-kinetic theory of gases is superior to phenomenological thermodynamics not because it tells us about such unobservable 'entities' as microscopic Brownian particles and submicroscopic molecules, but because it deals more adequately with such observable phenomena as we encounter looking into a microscope at what is usually called ''Brownian motion.'' Among the theories that are most widely accepted as empirically adequate today are many that make putative reference to unobservables, but—van Fraassen maintains—we can afford to remain agnostic about the actual existence of unobservables and judge theories on the basis of the degree to which they "save the phenomena."

Empiricism has traditionally come in two forms, *concept empiricism* and *statement empiricism*. Because of their view that all of our ideas are based upon sense experience, the 'British empiricists' became the most celebrated advocates of concept empiricism. Hume is the clearest case; in requiring that the meanings of concepts be analyzed in terms of the impressions from which they are derived, he formulated a criterion of cognitive meaning for concept empiricism. In the present century, this type of empiricism has found expression in operationism.

Twentieth-century logical empiricists, in contrast, have generally supported some form of statement empiricism. This view is often embodied in the verifiability criterion of cognitive meaning, which accords cognitive meaning to statements if and only if they can, in principle, be supported or undermined by observational evidence. There is no need, however, to become involved in the vexed question of meaning criteria (and I intend to leave that issue entirely aside). Statement empiricism can be formulated appropriately in the claim that all evidential support for factual statements about the world must, at bottom, be observational in character, and that formulation will be quite adequate for our purposes.

A fundamental question that naturally arises is whether these two forms of empiricism are equivalent to one another. I do not believe that they are. The term (or concept) "molecule" is not, in my opinion—and in the opinions of the majority of contemporary philosophers of science, I think—fully analyzable in terms of a purely observational vocabulary. Such concepts cannot be defined in a purely operational manner. Thus, it seems to me, the kinetic-molecular theory of gases cannot be accepted as literally meaningful and correct by a concept empiricist.

In this chapter I have been arguing that the experimental evidence for the existence of molecules and atoms that early twentieth-century scientists found compelling was, indeed, logically sound. If various statements about molecules—for example, that 4 grams of He^4 contain about 6×10^{23} monatomic molecules—are construed in any straightforward manner, they imply that such molecules exist. Whoever accepts such statements as *true* is committed to theoretical realism. In claiming that such statements can be established on the basis of experimental evidence, I have been adopting a statement empiricist stance. Without advocating any sort of meaning criterion, I am maintaining that statements about unobservables can be supported by empirical evidence. Whether this view is defensible depends upon the answer to a rather abstract question, namely, whether it is possible to have observational evidence for statements that cannot be fully formulated in observational terms. If an affirmative answer can be justified, then theoretical realism need not be incompatible with statement empiricism. This strikes me as a momentous issue. I should like to believe that an honest empiricist can have good grounds for believing that atoms, molecules, and perhaps even quarks exist—not merely that theories which seem to mention such 'entities' are empirically adequate.

Although it would be inaccurate to classify van Fraassen as a concept empiricist,[7] his constructive empiricism does exhibit some strong affinities to the traditional concept empiricism of Locke, Berkeley, and Hume. He is not literally a concept empiricist because his entire discussion is framed in terms of acceptance of and belief in statements. At the same time, he asserts that "empiricism requires theories only to give a true account *of what is observable*, counting further postulated structure as a means to that end" (1980, p. 3, italics in original). This thesis, which is reiterated in various ways throughout the book, seems to imply that the only statements that must face the question of truth or falsity are those that can be stated in observational terms. Indeed, in explaining the rationale for naming his own position, he says, "I choose the adjective 'constructive' to indicate my view that scientific activity is one of construction rather than discovery: construction of models that must be adequate to the phenomena, and not discovery of truth concerning the unobservable" (ibid., p. 5). It therefore appears that van Fraassen's constructive empiricism is equivalent to some form of concept empiricism.

To ask whether it is possible to have empirical evidence for statements about unobservables is to ask whether, from statements that can be established on the basis of observation, it is possible legitimately to infer statements about unobservables. It seems evident that no such feat would be possible if deductive inferences were the only admissible kinds. The basic issue seems to devolve upon the question of whether inductive logic contains the resources upon which such inferences can be founded. This is the question, I am claiming, that van Fraassen fails to confront directly.

Let us assume, with the constructive empiricist, that we can make at least a rough distinction between what is observable and what is not. Even if no sharp boundary can be drawn, there are clear cases of each type: automobiles and coffee cups are observable; electrons and quarks are not. Let us suppose, further, that experience has provided us with many actual observations, and that induction by enumeration has enabled us to establish a reasonable supply of empirical generalizations—that is, universal or statistical statements about regularities that hold among observables. The question is whether this empirical basis constitutes an adequate foundation

---

[7] In private correspondence, van Fraassen has informed me that he would consider it incorrect to classify his view as concept empiricism.

for inductive inferences to statements about unobservables. I shall try to sketch some grounds for an affirmative answer to this question.

There is a great deal of truth in Hume's remark (*An Inquiry Concerning Human Understanding*, sec. 4): "All reasonings concerning matters of fact seem to be founded on the relation of *cause and effect*. By means of that relation alone we can go beyond the evidence of our memory and senses." However, there does seem to be one important exception, namely, induction by simple enumeration. Induction by enumeration, as just suggested, yields empirical generalizations. These generalizations are not necessarily causal generalizations; they are the sorts of regularities for which we seek causal explanations. It also plays a crucial role in the reasoning used to extend our sensory apparatus by means of instruments. Consider a telescope. One can observe, with the naked eye, a distant object that appears small and lacking in detailed features. When it is observed with the aid of a telescope, it appears larger and more fully endowed with features. Relinquishing the telescope, we can approach the distant object and verify the fact that the telescope does indeed furnish us with observational information about the object not available by means of the unaided senses from that vantage point. Such generalizations about the properties of telescopes are completely empirical; they compare the observable properties of objects seen from different vantage points. Even the constructive empiricist can conclude, on the basis of this reasoning, that there are objects, such as the moons of Jupiter, that are too small and too distant to be observed from the earth by means of unaided human senses. Because the moons of Jupiter would be readily observable to an appropriately situated human observer, we have not stepped outside of the realm of entities whose existence is unproblematic for the constructive empiricist.

A similar approach can be taken toward magnifying glasses. One can open the compact edition of the *Oxford English Dictionary* and observe some words printed in type large enough to be read easily with the naked eye. One can see words in smaller type that can be read, but only with considerable difficulty, without the aid of a magnifying glass. When the magnifying glass is used, these words also are easy to read. Other words, printed in still smaller type, appear to the naked eye to consist of letters, but the letters cannot be discerned clearly. With the aid of the magnifying glass, even these words can be read, though not necessarily with ease. Similar experiments, using low-power microscopes to view small visible objects, can be conducted. Such experiments with telescopes, magnifying glasses, and microscopes provide a satisfactory basis for empirical generalizations about the optical properties of lenses.

There is, on van Fraassen's view, a fundamental difference between microscopes and telescopes. As I have already remarked, objects that are too far from earth to be seen with the naked eye by earthbound observers are not unobservable in principle, for they can be seen by normal human observers located elsewhere. Objects such as microscopic Brownian particles are not observable by observers with human sensory apparatus regardless of where they may be located; consequently, they lie outside of the realm of that which is observable in principle by human observers, and beyond the range of simple empirical generalization.

The next step in the argument is, of course, to make an inductive extrapolation from the foregoing generalizations about the optical properties of lenses used to observe visible objects to a full-blown theory of geometrical optics, from which we may infer that objects that are too small to be seen by the naked eye alone can be observed with the aid of microscopes.

How is this crucial step, which takes us from the realm of observables to that of unobservables, to be characterized? It is, I think, both a causal argument and an argument by analogy. The causal aspects of the situation are revealed by such considerations as the fact that light transmission is a causal process (as discussed in chapter 5) and the fact that intersections of light rays with lenses and other kinds of material objects are causal interactions (as discussed in chapter 6). Rather crudely speaking, the analogy takes the following form:

It is observed that:
An effect of type E_1 is produced by a cause of type C_1,
An effect of type E_2 is produced by a cause of type C_2,

.
.
.

An effect of type E_k occurred.

We conclude that:
A cause of type C_k produced E_k.

An analogical argument of this sort can take us from premises about observables to a conclusion about unobservables, for C_k may be an unobservable cause that is similar to C_1, C_2, . . . in most respects other than size. If this argument is logically correct—and I think it is—there can be inductive evidence for statements about observables. I thus agree with Cartwright (1983, p. 89) that "inference from effect to cause is legitimate." The legitimacy of this kind of argument implies that theoretical realism is compatible with statement empiricism. Indeed, as Cartwright (ibid., p.

160) further maintains, "All sorts of unobservable things are at work in the world, and even if we want to predict only observable outcomes, we will still have to look at their unobservable causes to get the right answer."

What is the logical status of analogical arguments? The answer one gives will, of course, depend upon the kind of inductive logic one adopts. In (Salmon, 1967, especially chap. 7), I elaborate in some detail what might be called an *objective Bayesian* approach. In a system of that sort, induction by enumeration is the primitive inductive rule, but all of the formal relationships furnished by the mathematical calculus of probabilities are available as well. Since Bayes's theorem is derivable within the probability calculus, it can be used to schematize more sophisticated inductive methods. Bayes's theorem cannot be applied, however, unless the prior probabilities it incorporates are supplied. One of the most important uses of analogies is, in my opinion, to provide prior probabilities. Because of the 'swamping of priors' by direct observational evidence, the numerical values of prior probabilities need not be known with much precision at all; what is needed is merely some sort of vague plausibility consideration (see Salmon, 1967, pp. 122, 129). Analogical arguments are frequently employed in that fashion; a few important examples illustrate the point. In physics, the inverse square character of the forces constitutes a strong analogy between Newton's gravitational theory and Coulomb's theory of electrostatics. In biomedical research, similarities between humans and other mammals provide the analogy that makes animal experimentation relevant to human physiology. In anthropology, observations of extant primitive societies are useful in interpreting the artifacts left by prehistoric cultures.

In the case at hand—the status of microscopic entities too small to be seen by human observers without the aid of instruments—the analogies are, when viewed schematically, quite simple and straightforward. Suppose that we observe a number of objects of different sizes under a particular low-powered microscope. An object that is 0.5 mm in diameter (as measured without using any optical instruments) appears under this microscope to have the same diameter as an object 2 mm in diameter viewed from the same distance without the aid of the microscope. An object that is 0.1 mm in diameter appears under the microscope to have the same diameter as an object that is 0.4 mm in diameter viewed from the same distance without the aid of a microscope. And so on. Suppose that no particle that I can see without the aid of a microscope has a diameter as small as 0.01 mm, but that I can see one whose diameter is 0.04. Then, if under the microscope I see a particle that appears to have the same diameter as one whose diameter is 0.04 mm as seen without the microscope, I infer by analogy that there is a particle in the field of the microscope that is too small to be seen with the naked eye. This analogy may not be a very strong inductive argument, but it does at least lend plausibility to its conclusion. The supposition that there is such a microscopic particle can be tested; for example, we can predict what will be seen when the same slide is viewed under microscopes of different types (see Hacking, 1981) and with different powers of magnification. Given only a nonzero value for the prior probability of the hypothesis that there is such a microscopic particle that produces the image, Bayes's theorem guarantees that a positive result of an observational test will inevitably enhance the probability of that hypothesis.

The inference, which I have just sketched in a highly oversimplified fashion, is discussed in concrete and fascinating detail by Ian Hacking in "Do We See Through a Microscope?" (1981). In considering the observation of dense bodies in red blood cells, Hacking mentions a device known as a microscopic grid. "Slices of red blood cell are fixed upon a microscopic grid. This is literally a grid: when seen through a microscope one sees a grid each of whose squares is labelled with a capital letter" (p. 315). He then proceeds to use the grid to address the issue van Fraassen has raised:

I now venture a philosopher's aside on the topic of scientific realism. Van Fraassen says we can see through a telescope because although we need the telescope to see the moons of Jupiter when we are positioned on earth, we could go out there and look at the moons with the naked eye. Perhaps that fantasy is close to fulfillment, but it is still science fiction. The microscopist avoids fantasy. Instead of flying to Jupiter he shrinks the visible world. Consider the grid that we used for re-identifying dense bodies. The tiny grids are made of metal; they are barely visible to the naked eye. They are made by drawing a very large grid with pen and ink. Letters are neatly inscribed by a draftsman at the corner of each square on the grid. Then the grid is reduced photographically. Using what are now standard techniques, metal is deposited on the resulting micrograph. . . . The procedures for making such grids are entirely well understood, and as reliable as any other high quality mass production system.

In short, rather than disporting ourselves to Jupiter in an imaginary space ship, we are routinely shrinking a grid. Then we look at the tiny disc through almost any kind of microscope and see exactly the same shapes and letters as were drawn in the large by the first draftsman. It is impossible seriously to entertain the thought that the minute disc, which I am holding by a pair of tweezers, does not in fact have the structure of a labelled grid. I know that what I see through the microscope is veridical because we *made* the grid to be just that way. (Pp. 316–317)

When we look at the methods actually employed by microscopists, we see that an impressively strong analogical argument can be extracted.

My argument for theoretical realism in this chapter has relied heavily upon the principle of the common cause. This principle may likewise be viewed as an appeal to analogy. As I tried to show in chapter 6, we have all had considerable success in explaining otherwise improbable coincidences by finding indirect causal connections via common causes (of either the conjunctive or the interactive variety). On this basis, we can say that the prior probability of the hypothesis that posits a common cause, given an otherwise improbable coincidence, is not negligible. In any two situations in which a conjunctive common cause is present, there is an analogy of causal structure. The same may be said of any two cases in which we find interactive forks. Moreover, the causal hypothesis can be confirmed by making experimental tests.

Hacking's example of the microscopic grid furnishes an excellent illustration. In spite of some disclaimers elsewhere in his article, he proceeds immediately to invoke a common cause argument strikingly similar to that which I have used earlier in this chapter:

> I know that the process of manufacture [of the grid] is reliable, because we can check the results with the microscope. Moreover, we can check the results with any kind of microscope, using any of a dozen unrelated physical processes to produce an image. Can we entertain the possibility that, all the same, this is some gigantic coincidence[?] Is it false that the disc is, in fine, in the shape of a labelled grid? Is it a gigantic conspiracy of 13 totally unrelated physical processes [recall Perrin's table of 13 ways of ascertaining Avogadro's number] that the large scale grid was shrunk into some non-grid which when viewed using 12 different kinds of microscopes still looks like a grid? (1981, p. 317)

Arguments of these sorts provide strong inductive warrant for the conclusion that there are entities too small to be detected with the naked eye— including such things as Brownian particles—whose existence and behavior can be ascertained by means of the microscope.

Having crossed the boundary set by the constructive empiricist by admitting the existence of microscopic entities, we can use arguments of the same logical kind to descend to objects of submicroscopic scale. Consider once more the ascertainment of Avogadro's number. Suppose that we look at Rutherford's work on alpha radiation and helium production and at Perrin's work on Brownian motion. I claimed previously that we could interpret the situation in terms of a conjunctive fork, and that the evidence supports the hypothesis that there are about 6×10^{23} molecules in a mole of any given substance. Taking account of the relationships among molecules, atoms, electrons, and ions, we could predict that the outcome of an experiment on the electrolysis of silver would yield the same value for the number of electron charges in a faraday. A positive outcome of the experiment would strongly support the hypothesis that there are indeed that many silver ions in a mole of silver. A fortiori, silver ions exist. Additional experimental tests using other methods of ascertainment of Avogadro's number provide further confirmation. It will be recalled that Perrin mentioned three distinct methods of ascertainment in his 1908 papers, and thirteen distinct methods in his 1913 book. If there were some alternative hypothesis with even a modicum of plausibility to account for the amazing agreement, these results might fail to provide strong support for the existence of molecules and other micro-entities. As I tried to show, no such reasonable alternative seems to be available.

It should be noted that the principle of the common cause, when approached as an analogy that has a bearing upon prior probabilities, makes no pretense of being a universal or immutable principle of inference. In the domains in which it has been successful, it confers initial plausibility upon causal hypotheses that appeal to common causes. In the quantum domain, as we shall see in greater detail in the next chapter, there are serious grounds for doubt about its applicability. That fact does not undermine its utility in dealing with other kinds of natural phenomena.

There is another way of describing the same feature of the common cause principle. The basic form of inductive reasoning can be applied— at a higher level, so to speak—to evaluate other methods of nondemonstrative inference. If a given principle of inference has proved to be quite reliable in past applications, then we may reasonably have some confidence in further uses of it. This is, I suppose, the fundamental principle underlying legitimate appeals to authority. A reliable authority is one that has a good record of providing accurate information. The basic point can be stated more generally: secondary methods of ampliative inference can be validated in terms of their past successes by appeal to the primary inductive method. In domains outside of quantum physics, the principle of the common cause has an impressive record.

Although, as I have said, I do not intend to argue the case for physicalism as opposed to phenomenalism, it does seem worthwhile to mention the fact that the common cause principle has an important bearing upon that issue. A significant feature of our knowledge of the world in which we live is that it comes to us through more than one modality of sense. Visual information about the shape of a physical object can be reinforced by touch, and rechecked by reference to what we hear other observers say about it. This is the argument for epistemological realism that Popper

(1972, pp. 42–43) attributes, with great approbation, to Sir Winston Churchill. Churchill's argument clearly supports theoretical realism as well.

The fundamental difference between van Fraassen's theory of scientific explanation and that which I have been advocating relates closely to our disagreement regarding theoretical realism. If one adopts an epistemic approach to scientific explanation—whether it be the inferential, information-theoretic, or erotetic version—success in providing scientific explanations requires of explanatory theories only that they be empirically adequate. These theories enable us to construct the deductive or inductive arguments that are taken by advocates of the inferential version to constitute scientific explanations. Such theories can be viewed by the proponent of the information-theoretic version as transmitting information in the manner required for that approach. Theories of the same sort can be used to provide answers to explanation-seeking why-questions that are considered adequate from the standpoint of van Fraassen's erotetic version. When we accept such theories for purposes of explanation, according to van Fraassen, we need only maintain that they are empirically adequate, not that they provide true descriptions of unobservable 'entities.' The epistemic approach allows us to remain agnostic regarding the existence of unobservables.

It would be altogether contrary to the spirit of the ontic approach—concerned, as it is, with the causal mechanisms that produce the facts-to-be-explained—to allow agnosticism of that sort. If the account of causal processes, causal interactions, and causal forks offered in preceding chapters is anywhere near correct, then causal mechanisms frequently involve unobservable entities. Consider, for example, the causal interactions involved in Brownian motion. According to Einstein's theoretical account of this phenomenon, the microscopic particle undergoes many collisions with the molecules of the fluid in which it is suspended. If there are no such things as molecules, then we have a radically inaccurate account of the mechanism. If we nevertheless accept Einstein's theory as empirically adequate, and use it to explain Brownian motion, then it seems to me that we come uncomfortably close to Hans Vaihinger's philosophy of 'as if' (Vaihinger, 1924). The Brownian particle behaves as if it is being jostled by myriad submicroscopic molecules. On van Fraassen's view, of course, we cannot say that there are no such things as molecules, so they cannot be considered merely fictitious 'entities,' but we cannot say that they actually exist. For the ontic approach, any causal mechanism that is invoked for explanatory purposes must be taken to be real. If we are not prepared to assert its existence, we cannot attribute explanatory force either to that mechanism or to any theory that involves it. As Cartwright (1983, essay 5) argues cogently, fictitious causal mechanisms do not have explanatory import. The tooth fairy does not explain anything.

9 | The Mechanical Philosophy

A BASIC THEME of the present book has been to draw a sharp contrast between what I have called the *epistemic conception* and the *ontic conception* of scientific explanation. I have attempted to subject the epistemic conception—which, in its inferential version, has been the dominant conception for several decades—to severe criticisms, and to advocate adoption of the ontic conception as a more viable alternative. In order to elaborate the ontic conception, it has been necessary to develop a theory of causality that departs fundamentally from standard approaches. The theory here proposed appeals to causal forks and causal processes; these are, if I am right, the mechanisms of causal production and causal propagation that operate in our universe. These mechanisms, it has been emphasized, may operate in ineluctably stochastic ways. Scientific explanation, according to the ontic conception, consists in exhibiting the phenomena-to-be-explained as occupying their places in the patterns and regularities which structure the world. Causal relations lie at the foundations of these patterns and regularities; consequently, the ontic conception has been elaborated as a *causal conception* of scientific explanation. As I suggested in chapter 1, the causal conception accords harmoniously with at least some of our basic intuitions about the nature of scientific understanding.

Logic versus Mechanisms

One striking difference between the ontic conception and the epistemic conception (in its time-honored inferential version, at least) pertains to their claims of universality. Since this version of the epistemic conception takes scientific explanations to be arguments, it is natural for philosophers to attempt to characterize 'the logic' of explanation; indeed, the landmark 1948 essay by Hempel and Oppenheim was entitled "Studies in the Logic of Explanation," and it endeavored to provide a semantic explication of one pattern of scientific explanation. In subsequent works, Hempel offered semantic treatments of other patterns. There is, of course, no suggestion in any of these discussions of the logic of scientific explanation that all phenomena in this world are necessarily amenable to explanation—let alone that phenomena in all possible worlds must be explainable. However, the strong suggestion does seem present that insofar as phenomena in this or any other possible world can be explained, the explanations must conform to one or another of the logical patterns that have been elaborated. It is my view that attempting to give a logical characterization of scientific explanation is a futile venture, and that little of significance can be said about scientific explanation in purely syntactical or semantical terms. I believe, rather, that what constitutes adequate explanation depends crucially upon the mechanisms that operate in our world. In all of this there is—obviously—no logical necessity whatever.

The clearest illustration of this point lies, perhaps, in the principle of the common cause. With respect to conjunctive forks, for example, there is no logical guarantee that conjunctive common causes must exist to explain what would otherwise be improbable coincidences, nor that, if they exist, they must be temporally prior to their effects rather than subsequent to them. Nevertheless, in our world—at least outside of the quantum domain—the use of conjunctive common causes to explain statistical correlations has been remarkably successful. In view of such success, we would be reluctant to relinquish this explanatory principle. If, however, we lived in a world that had a radically different causal structure from our own, we might find other explanatory principles useful and appealing.

I have tried to say something constructive about some of the mechanisms that function in certain domains of our world; in other worlds, or in other domains of this world, mechanisms of a radically different sort might operate. I have focused upon causal mechanisms; however, mechanisms of a noncausal sort might be found. I have tried to describe certain kinds of causal mechanisms that happen to exist in our world; perhaps causality can take different forms.[1] I make no claim for universal applicability of my characterization of scientific explanation in all domains of our world, let alone for universality across all possible worlds. Thus I agree with Achinstein (1983, chap. 4) in rejecting the universalist viewpoint regarding scientific explanation. But I do maintain that scientific explanation is designed to provide understanding, and such understanding results from *knowing how things work*. In this sense, the theory of scientific explanation that I

---

[1] In his address to the 1982 meeting of the Philosophy of Science Association, Howard Stein expressed a similar idea in the following terms: "I think that we—perhaps physicists, and certainly philosophers—too often fall into the trap of presuming that we have an adequately clear concept of what is called 'causation.' I have never considered Hume's analysis of the notion of 'cause' a satisfactory one; but I think that Hume was quite right in maintaining that whatever we know of causes—or, as I should prefer to say, of causal nexus, or interconnection, or interaction—we know not *a priori* but only on the basis of experience; and I should add that our best such knowledge is contained precisely in our best scientific theories."

have been attempting to develop is an expression of a *mechanical philosophy*.[2]

What is often characterized as a mechanistic world view is not very popular at present, largely because it is usually conceived in terms that are scientifically outmoded. A good—if somewhat primitive—example of a mechanistic world picture is the atomism of Lucretius, Gassendi, or Laplace. In the latter part of the nineteenth century, mechanism was identified with the view of those physicists who tried to explain electromagnetic phenomena in terms of a mechanical ether. This viewpoint is epitomized in the famous remark of Lord Kelvin (1884, p. 270): "I never satisfy myself until I can make a mechanical model of a thing. If I can make a mechanical model I can understand it." Thus mechanism is often identified with the notion that explanations of physical phenomena are inadequate unless they are given in terms of levers, springs, pulleys, strings, wheels, gears, deformable jelly, and so forth. Construed in these terms, mechanism *is* scientifically anachronistic.

The difficulty with this late nineteenth-century version of the mechanical philosophy lies not in the fundamental philosophical perspective, but, rather, in its misconception of the *basic* mechanisms that actually operate in the physical world. We realize today that the mechanical properties of such material objects as springs and gears require explanation in terms of the fine structure of macroscopic objects. We recognize, for example, that electrostatic forces play a crucial part in explaining such structures. Because of Maxwell's electromagnetic theory and Einstein's special theory of relativity, we now believe that the electromagnetic field has fundamental physical reality. Electromagnetic phenomena are not to be explained in terms of the behavior of a mechanical medium. The situation is just the reverse. The mechanical properties of the macroscopic objects used by Lord Kelvin to make his models are to be explained in terms of the properties of fields and of the waves that propagate through such fields. As I said in the discussion of causal processes in chapter 5, the propagation of such waves through these fields is a fundamental causal mechanism in our universe. We have to change our mechanistic view from the crude atomism that recognizes only the motions of material particles in the void to a conception that admits such nonmaterial entities as fields, but for all of that, it is still a mechanistic world view. Materialism is untenable, but the mechanical philosophy, I believe, remains viable.

Another contingent feature of my account of scientific explanation is the emphasis I have placed upon spatiotemporally continuous causal con-

---

[2] See (Westfall, 1971, chap. 2) for an exposition of this general viewpoint in its historical context.

nections, and my rejection of action-at-a-distance. As long as we remain outside of the quantum domain, I believe this demand is reasonable, although one argument to the contrary must be considered (see Glymour, 1982). According to this argument, there were, prior to the middle of the nineteenth century, two fundamental physical theories, each of which had enormous explanatory power, that incorporated action-at-a-distance. One of these was Newton's theory of gravitation and the other was Coulomb's theory of electrostatics. Even though many people, including Newton himself, were uncomfortable with the idea of action-at-a-distance, it would be wrongheaded—so the argument goes—to rule out on a priori philosophical grounds explanations that appeal to these theories.

Speaking from a historical standpoint, I am in complete agreement with this line of thought. Before the beginning of the present century, it seems to me, our understanding of the basic mechanisms of nature did not preclude action-at-a-distance. As I have said repeatedly, I have no intention of trying to impose a priori constraints upon the nature of scientific explanation. Nevertheless, by the end of the nineteenth century, the success of Maxwell's electromagnetic field theory, along with Hertz's experimental detection of electromagnetic waves, showed that action-at-a-distance is not required in dealing with electromagnetic phenomena. Electromagnetic influences are propagated in a spatiotemporally continuous fashion at a finite velocity through electromagnetic fields. In the relativistic gravitational theories based upon Einstein's theory, similar considerations apply to the propagation of gravitational influences. Gravitational fields have physical reality, and gravitational influences are propagated as waves traveling at a finite velocity through such fields. Although—in contrast to electromagnetic waves—gravitational waves have not been directly detected as yet, there seems, nevertheless, to be sufficient indirect evidence to justify our believing that they exist (see Davies, 1980). Although appeal to action-at-a-distance might not have been an obstacle to the adequacy of a scientific explanation at earlier times, the situation has dramatically changed as a result of the successes of the theories of Maxwell and Einstein. At least, this attitude seems well-founded as long as we steer clear of the quantum domain. At the macroscopic level, action-by-contact seems to be a legitimate demand, but in quantum mechanics, the question of action-at-a-distance raises profound difficulties.

EXPLANATION IN QUANTUM MECHANICS

Although there may be differences of opinion on the question of whether quantum mechanics is compatible with determinism, I think it must be admitted that quantum theory forces us at least to consider seriously the possibility that indeterminism is true. Throughout this book, I have insisted that our philosophical theories of scientific explanation must take account of the *possibility* that some events are irreducibly stochastic occurrences, which are not completely determined in all of their characteristics by preceding causes. From the traditional standpoint, causality is incompatible with indeterminism—that is, to whatever extent and in whatever respects events are not completely determined, to that extent they cannot be explained causally. I have taken a different stance. According to the explicitly probabilistic account of causality developed in preceding chapters, it is possible to provide *causal* explanations of ineluctably stochastic events. Even if one adopts the received statistical interpretation of quantum theory, its indeterministic character poses no obstacles to the causal account of explanation I have been advocating.

The quantum theory developed by Heisenberg, Schroedinger, et al., in the mid-1920s does, however, pose other profound difficulties concerning causality. Reichenbach (1944) characterized the problem in terms of his *principle of anomaly*, and Bohr introduced his well-known *principle of complementarity* to deal with it. Let us consider a simple standard example—the Young two-slit experiment. If light of a given wavelength impinges upon a screen with one slit of suitable width, the light that passes through will exhibit a well-known diffraction pattern when it falls upon a second screen. If light of the same wavelength impinges upon a screen with two such slits, it will exhibit a characteristic two-slit diffraction pattern. The diffraction pattern that results when both slits are open simultaneously is radically different from the pattern that would result if the two slits were opened one at a time and the two one-slit diffraction patterns were superimposed. This fact about light has been known since the beginning of the nineteenth century, and it was taken as strong confirmation for the wave theory of light. The wave theory explains the difference between the two-slit diffraction pattern and the superposition of two one-slit patterns in terms of the mutual interference of the waves emanating from the two slits when they are open simultaneously. As long as light is regarded as a wave, this experiment poses no difficulties.

Phenomena such as the Compton scattering of X rays by electrons, however, suggest that light sometimes behaves more like a corpuscle than a wave. Similarly, electrons—which behave like particles in such contexts as Compton scattering—exhibit characteristic wave behavior under other experimental circumstances such as the Davisson-Germer diffraction of electrons by a nickle crystal. Taken together, these phenomena lead to the conclusion that both light and material particles exhibit a so-called wave-particle duality. If a Young two-slit apparatus with slits of the order of one angstrom in width could be set up, then we should expect exactly the same sort of two-slit interference pattern with electrons as was in fact observed with light. We may, therefore, speak of a Young two-slit *thought* experiment with electrons. This thought experiment has been widely discussed in the literature (see, e.g., Feynman et al., 1965, chap. 1), and there is basic agreement about its essential features.

The fundamental problem of the wave-particle duality can be seen in the following way. If we set up the two-slit apparatus and allow a large number of electrons to pass through the slits, the overall pattern will be just the sort of diffraction pattern we should expect if electrons are waves. If, however, we use a relatively weak source of electrons so that the arrival of each electron can be detected separately, we find that they arrive in a highly localized fashion, just like well-behaved corpuscles. The problem is this. If electrons are particles, then the two-slit pattern should be the same as the superposition of two one-slit patterns. Otherwise, it seems, an electron passing through one of the slits has to 'know' whether the other slit is open or closed. This appears to be a case of action-at-a-distance; the behavior of a particle passing through one slit seems to be affected by the state of the other slit. This is the sort of phenomenon to which Reichenbach attached the label "causal anomaly." We escape this causal anomaly if we treat the electron as a wave rather than a particle. If, however, the electron is a wave, then when it arrives at the second screen it should be spread out over a large area, and it should not collapse instantaneously into a pointlike region. As an analogy, imagine the surf arriving in waves spread out along Australia's Ninety Mile Beach. When each wave arrives, however, it suddenly collapses into a small bundle, depositing all of its energy at a single place. A bather who had the misfortune to be standing at such a point might land in Canberra (about 200 miles away) as a result. Hence, returning to our thought experiment, if we regard the electron as a wave, then its instantaneous collapse upon detection is another case of action-at-a-distance; parts of the wave remote from the detector influence the localized detector. Again, we have one of Reichenbach's causal anomalies. We can escape this anomaly by going back to the particle interpretation of the electron. Reichenbach maintained that by judicious choice between the wave and particle interpretations, it is possible to eliminate any given anomaly, but there is no single interpretation that can eliminate all causal anomalies from such a quantum mechanical system.

Bohr dealt with problems of this sort by invoking his principle of complementarity. The wave description and the particle description of the two-slit experiment are mutually incompatible descriptions of the same quantum mechanical situation. To apply both simultaneously would lead to self-contradiction. Nevertheless, Bohr claimed, each is a correct description of one aspect of the system, but the system as a whole cannot be understood entirely in terms of any single description. In one particular formulation of his principle of complementarity, Bohr claimed that the 'space-time description' and the 'causal description' of a quantum mechanical system are complementary. This implies that there cannot be a single consistent description that explains what happens in terms of spatiotemporally continuous causal processes and local causal interactions. Considerations of this sort seem to suggest that the theory of causal explanation that I have tried to develop in preceding chapters simply cannot be extended into the quantum domain.

A suspicion has sometimes been voiced that the foregoing difficulties concerning the two-slit experiment arise from an illicit use of such classical concepts as 'wave' and 'particle' in the quantum mechanical context. Even if it were possible to resolve these problems by eschewing improper use of classical concepts—which I seriously doubt—we would still have to come to terms with issues concerning action-at-a-distance and causal anomalies that were raised in the classic paper by Einstein, Podolsky, and Rosen, "Can Quantum-Mechanical Description of Physical Reality Be Considered Complete?" (1935). The profound problems raised in this paper continue to perplex thoughtful interpreters of quantum theory, and they have a direct bearing upon the issues concerning scientific explanation that we have been discussing.

The EPR problem, as it was presented in the original paper, concerned a composite physical system S composed of two parts, S_1 and S_2. At the outset, the two parts of S are in direct physical interaction with one another, but they are subsequently separated, and from that point on they do not interact with one another. The entire system S is in a state such that, even after the parts are separated, if a precise measurement is made that reveals the exact position of S_1, then one can calculate the exact position of S_2, and if a precise measurement is made that reveals the exact momentum of S_1, then one can calculate the exact momentum of S_2. The composite system S evolves in such a way that its *total* momentum remains constant, and the *relative* separation between the parts always has an exact value. Now, according to the quantum theory, as it is usually understood, the subsystem S_2 does not simultaneously possess both an exact position and an exact momentum. However, as Einstein, Podolsky, and Rosen point out, without disturbing or interacting with the subsystem S_2 in any way, we can ascertain either an exact position or an exact momentum of S_2 by measurements made on S_1. Moreover, by the time the measurement is made on S_1, the separation between S_1 and S_2 could be spacelike. Hence, according to special relativity, a measurement made on S_1 could not possibly have any effect on S_2. Einstein, Podolsky, and Rosen concluded that—quantum theory to the contrary, notwithstanding—the system S_2 does possess simultaneously both an exact position and an exact momentum.

More recent treatments of the EPR problem have often reformulated it in a somewhat simpler manner as suggested by David Bohm (1951, pp. 614–615). Bohm discusses a diatomic molecule, but we can take a still simpler example. Consider as our composite system S an atom of positronium—that is, a positron (positive electron) S_1 and a negative electron S_2 orbiting about one another. Let the total intrinsic angular momentum (spin) of the system be zero. Now let the two particles be separated from one another by some means (such as microwave radiation) that does not affect the angular momentum of the total system or of either part. Suppose that the two subsystems are moving apart along the x axis. After they have been separated, we can decide to measure (by means of a Stern-Gerlach apparatus) the y component of spin of the positron; the result will be either the definite value $+1/2$ or the definite value $-1/2$. Under these conditions, the y component of spin of the electron must have either the definite value $-1/2$ or the definite value $+1/2$, respectively. In saying that each particle has a *definite* value of spin in the y direction, I mean that each one is in an eigenstate of the y component of spin. I shall say more about the significance of this assertion later.

Alternatively, after the two particles have been separated, we can decide to measure the z component of spin of the positron. Again, the result will be either the definite value $+1/2$ or the definite value $-1/2$. Under *these* conditions, the z component of spin of the electron must have either the definite value $-1/2$ or the definite value $+1/2$, respectively. In *this* case, each particle is in an eigenstate with respect to the z component of spin. However, if the electron is in an eigenstate of the y component of spin, it *cannot* be in an eigenstate of the z component; and if it is in an eigenstate of the z component of spin, it *cannot* be in an eigenstate of the y component.[3]

If the electron is in an eigenstate (say $+1/2$) of the y component of spin, then a subsequent measurement of that component of spin will necessarily yield the same result. This means, more precisely, that if we prepare a large number of electrons in the eigenstate $+1/2$ of the y component of spin, then the result of a subsequent measurement of the y component of spin yields the value $+1/2$ *for each member of the ensemble*. If, however, we prepare the ensemble in the eigenstate $+1/2$ of the z component of spin, then subsequent measurements of the y component of spin will yield a random sequence of $+1/2$ and $-1/2$ results. This is what

[3] If a particle is in an eigenstate with respect to the z component of spin, it is in a superposition of the states $+1/2$ and $-1/2$ with respect to the y component, and conversely.

is meant by saying—as I did previously—that an electron that is in an eigenstate of spin in the z direction cannot be in an eigenstate of spin in the y direction. Mutatis mutandis, an electron in an eigenstate of spin in the y direction cannot be in an eigenstate of spin in the z direction. The EPR problem is this. A measurement performed upon the positron seems to influence the physical state of the electron, even though there can be no physical interaction between the two particles at the time the measurement is made. Indeed, the actual decision as to what measurement will be performed upon the positron can be made after the physical interaction between the two particles has ceased.

Let us attempt to see how the idealized EPR thought experiment might be conducted. Suppose that we have a large number of positronium atoms, all prepared in the same physical state, and that all have a total intrinsic angular momentum of zero. In each case, we separate the positron from the electron. *After* the separation has occurred, we use a table of random numbers to decide whether to make a measurement upon the positron or the electron. Then, we use another random device to determine whether to measure the y component or the z component of the spin of the particle in question. In each case, if we decide to make a measurement on the positron, we do not make any measurement on the corresponding electron, and conversely. In either case, if we decide to measure the y component, we make no attempt to measure the z component, and conversely. After a long series of such measurements has been made, we examine the results. We look, for example, at all measurements made on the y component of spin of electrons. We find that the results of all of these measurements constitute a random sequence of + 1/2 and − 1/2. If we look at the sequence of all measurements made on the z component of spin of positrons, again we find a random sequence of + 1/2 and − 1/2. The same goes for the measurements of the z component for electrons, and also for the y component for positrons. When all of these results have been well established, we begin a new series of experiments.

We start again with a large number of positronium atoms, all in the same state and all with total intrinsic angular momentum of zero. Again, in each case, the electron and positron are gently separated. In this series of experiments, we use a random device to decide in each case (after the separation has been effected) whether to measure the y component or the z component, but this time we measure the component in question on both the electron and the positron. When we do so, we find a perfect negative correlation of results. For either component, if the value for one subsystem is + 1/2, the value for the other is − 1/2. This astonishing correlation demands explanation—a point ignored by Heinz Pagels (1982, pp. 151–152) in his dismissive treatment of this problem.

There is no doubt, I believe, that standard quantum theory enables us to derive the foregoing results (see Cantrell and Scully, 1978, for details). But this fact does not mean that we can provide a satisfactory explanation of the predicted experimental result. In the provocative article to which I alluded in chapter 1, Bernard d'Espagnat (1979) addresses precisely this issue:

> Any successful theory in the physical sciences is expected to make accurate predictions. Given some well-defined experiment, the theory should correctly specify the outcome or should at least assign the correct probabilities to all the possible outcomes. From this point of view quantum mechanics must be judged highly successful. As the fundamental modern theory of atoms, of molecules, of elementary particles, of electromagnetic radiation and of the solid state it supplies methods for calculating the results of experiments in all these realms.
>
> Apart from experimental confirmation, however, something more is generally demanded of a theory. It is expected not only to determine the results of an experiment but also to provide some understanding of the physical events that are presumed to underlie the observed results. In other words, the theory should not only give the position of a pointer on a dial but also explain why the pointer takes up that position. When one seeks information of this kind in the quantum theory, certain conceptual difficulties arise. For example, in quantum mechanics an elementary particle such as an electron is represented by the mathematical expression called a wave function, which often describes the electron as if it were smeared out over a large region of space.
>
> This representation is not in conflict with experiment; on the contrary, the wave function yields an accurate estimate of the probability that the electron will be found in any given place. When the electron is actually detected, however, it is never smeared out but always has a definite position. Hence it is not entirely clear what physical interpretation should be given to the wave function or what picture of the electron one should keep in mind. Because of ambiguities such as this many physicists find it most sensible to regard quantum mechanics as merely a set of rules that prescribe the outcome of experiments. According to this view the quantum theory is concerned only with observable phenomena (the observed position of the pointer) and not with any underlying physical state (the real position of the electron). (P. 158)

In citing the example of the electron, d'Espagnat is clearly referring to the problem of wave-particle duality mentioned previously. His main topic of concern is, however, experiments of the EPR variety, and the problem of providing adequate explanations of the quantum mechanical results.

d'Espagnat mentions three assumptions that lead to the problem concerning the explanation of the quantum mechanical results. The first assumption is *realism*—that is, the assumption that such entities as atoms and subatomic particles exist. I discussed the realistic thesis in the preceding chapter, and shall accept d'Espagnat's first assumption without further comment.

The second assumption is *induction*. This means that the observed correlations between the spin states of the electrons and positrons are stable statistical relations. The correlations within the observed sample are not a matter of mere accident. Suppose, for instance, that two players simultaneously toss two standard pennies in the standard manner a number of times, and that the results on the two coins are *in fact* statistically independent of one another. Even so, it would not be too surprising to discover that on two occasions, when the first coin landed heads the second landed tails and vice versa. Within the sample of two pairs of tosses, we have an accidental perfect negative correlation between the results of the tosses. It would be fairly surprising if the same perfect negative correlation persisted for ten observed pairs of tosses. It would be utterly astonishing if—given that the outcomes are *in fact* independent—a perfect negative correlation should be observed in several hundreds or thousands of tosses of the two coins. Indeed, under these circumstances, we would surely find ourselves driven to challenge—and probably reject—the assumption that the tosses on the two coins are actually statistically independent of one another. d'Espagnat's inductive assumption is simply the assumption that perfect negative (or positive) correlations among large numbers of cases are not mere chance coincidences, but, rather, they represent genuine statistical correlations. We shall accept this (methodological) assumption as well, for if we were to reject induction, there would be no way in which any scientific generalization could be established on the basis of empirical evidence. Having accepted the first two assumptions, we have agreed that there are certain kinds of micro-entities, and certain kinds of genuine statistical correlations in their behavior, as described in the foregoing example involving the spin states of electrons and positrons. The quantum theory predicts such experimental results, and the experimental evidence seems to bear out the predictions.

The problem we now face—which involves d'Espagnat's third assumption—is how to explain the quantum mechanical correlations. d'Espagnat articulates the issue in terms that are familiar from preceding chapters:

> Whenever a consistent correlation between such events is said to be understood, or to have nothing mysterious about it, the explanation offered always cites some link of causality. Either one event causes the

other or both events have a common cause. Until such a link has been discovered the mind cannot rest satisfied. Moreover, it cannot do so even if empirical rules for predicting future correlations are already known. A correlation between the tides and the motion of the moon was observed in antiquity, and rules were formulated for predicting future tides on the basis of past experience. The tides could not be said to be understood, however, until Newton introduced his theory of universal gravitation. (P. 160)

The question, in short, is whether we can give a *causal explanation* of the quantum mechanical result. d'Espagnat's treatment of this issue is particularly useful for our discussion because, unlike other formulations of the problem with which I am acquainted, he appeals explicitly to such causal mechanisms as common causes.

The third assumption d'Espagnat enunciates is called "Einstein separability" or "Einstein locality"—the principle which states that information or causal influence cannot be propagated at a velocity greater than the speed of light. This principle would be violated if there were tachyons, or other superluminal signals, which could be used to transmit information. More importantly, from our standpoint, the assumption of locality or separability precludes action-at-a-distance. Along with d'Espagnat, I am inclined to reject the hypothesis of superluminal signals. Although it may be true, as has sometimes been argued, that tachyons are compatible with special relativity, all attempts to detect them have had negative results. Moreover, if tachyons did exist, they would engender fundamental violations of causality (see Salmon, 1975, pp. 122–125). Thus, it seems to me, the serious question at this point concerns the issue of spatiotemporally continuous propagation of causal influence versus action-at-a-distance.

Theories that satisfy d'Espagnat's three assumptions of realism, induction, and locality are called *local realistic theories*. Using Bell's inequality, d'Espagnat shows that such theories make predictions that conflict with the predictions of standard quantum mechanics. Moreover, he points out, to the extent that empirical tests have been performed, the evidence seems to support quantum theory against the local realistic theories. Although the argument is not rigorous, the strong suggestion is that nature seems to exhibit action-at-a-distance in the quantum domain.

Let us go back and look at the EPR experiment on the spin states of positrons and electrons. While the two particles constitute a positronium atom, there is a direct physical interaction between them. The total intrinsic angular momentum of the system is zero. When the two particles move away from one another, neither is in an eigenstate of spin with respect to the z component, but the sum of the two spins with respect to the z

component has the definite value zero. If a measurement on the positron puts the positron into an eigenstate, say, spin up, then the electron is automatically forced into the eigenstate of spin down with respect to the z component. An analogous story would be told if the measurement of spin were made with respect to the y component. Since the electron cannot simultaneously be in an eigenstate with respect to the z axis and in an eigenstate with respect to the y axis, we must conclude that, in some cases at least, the measurement of a component of spin on the positron alters its spin state, and thus alters the spin state of the remote correlated electron.

It must be recalled, of course, that the positronium atom is a single physical system, and that this system remains a single system that can be described in terms of a single state function even after the parts have been spatially separated from one another. Thus a measurement of the spin of the positron is an interaction with the *entire* system, and the state of the *entire* system is modified as a result. This state describes the electron as well as the positron. If we look at the EPR experiment in these terms, we have a special case of the general problem of the 'collapse of the wave function' (or 'reduction of the wave packet'). Just as the localized detection of an electron described in terms of a wave that is spread out in space involves an instantaneous change of state over a large region, so, also, the spin measurement on the positron has an effect upon the state of a system that is also spread over a large spatial region. It is this aspect of quantum theory that is so deeply puzzling from a causal standpoint. When we ask for the causal (or other sort of) mechanism involved in the production of the EPR correlations, we find ourselves at a loss. If one adopts an ontic or causal conception of scientific explanation, the foregoing situation is profoundly disturbing.

The perplexity can, perhaps, be avoided by adopting a radical instrumentalist standpoint, or an old-fashioned positivism of the Mach-Pearson variety. On this sort of view, the only function of scientific theories is to articulate, in as simple and concise a manner as possible, empirical relationships among observables. To adopt this position involves abandoning all attempts to *explain* the empirical phenomena. The sole purpose of the scientific theory is, on this view, to facilitate prediction of observable occurrences.

As we saw in the preceding chapter, an attitude of this sort motivated some of the opponents of atomism in the nineteenth century. I argued at length that the experimental work on Avogadro's number in the early years of the twentieth century provided rather conclusive proof of the existence of atoms and molecules. It seems to me that the molecular-kinetic theory provides an admirable explanation of such empirical relationships as the ideal gas law. The ideal gas law, in contrast, seems to have little or no explanatory force in the absence of the kinetic-molecular theory. It would strike me as an undesirable and regressive move to retreat to instrumentalism in the face of the EPR problem.

Consider an example of an empirical law from the quantum domain. Like the ideal gas law, Balmer's empirical formula for the lines in the hydrogen spectrum has no explanatory value in and of itself. It does not *explain why* the emission spectrum of hydrogen contains a line at 4,340 angstroms.[4] Rather, the fact that there is a series of lines in the spectrum satisfying the equation

$$\lambda = b \times \frac{n^2}{n^2 - 2^2}$$

cries out to be explained. By appealing to the behavior of such subatomic entities as protons and electrons, it is possible to provide a quantum mechanical explanation of a great many features of the hydrogen spectrum. This mechanical explanation obviously departs in fundamental ways from the sort of mechanical explanation that would have seemed satisfying to Lord Kelvin.

Proponents of the inferential version of the epistemic conception of scientific explanation need not be troubled by the perplexities we have been discussing. According to that conception, to explain a phenomenon is to show that a statement of its occurrence follows from initial conditions and lawful generalizations. Quantum theory has been remarkably successful in enabling us to derive correct statements about observed phenomena, and, as mentioned previously, quantum theory provides the means to calculate the results of the EPR experiment. According to those who advocate the inferential approach, such 'explanations' are correct and legitimate. To some philosophers, this consideration might seem to constitute a strong argument in favor of the inferential conception. If we adopt the inferential approach to scientific explanation, it might be claimed, then we already have a quantum mechanical explanation of the EPR correlations. The inferential conception allows us to avoid the nasty problems we have been discussing. Since it would be pleasant to put them aside, such philosophers would advise us to embrace forthwith the inferential conception. To my mind, the moral of the story is just the reverse. Since our ability to calculate the results of experiments on correlated systems manifestly does not suffice to explain these correlations, the inferential conception of scientific explanation must be deficient.[5]

---

[4] Hempel (1966, p. 53) does appear to attribute some degree of explanatory power to Balmer's formula.

[5] Van Fraassen (1980, p. 122) addresses this issue as follows: "Salmon mentions explicitly the limitation of this account to macroscopic phenomena (though he does discuss Compton

It is not quite clear to me how van Fraassen's erotetic version of the epistemic conception is supposed to deal with explanation in quantum mechanics. It may be fair to say that pragmatic factors have overriding importance. What why-questions are asked, and what answers are considered satisfying, may depend heavily upon the context.

Near the beginning of his chapter on explanation, van Fraassen (1980, pp. 101–103) offers three "workaday examples of scientific explanation." It is interesting to note that, in each case, the explanation consists of a deduction of the explanandum from an explanans that contains at least one general law. The erotetic and inferential versions of the epistemic conception have strong affinities. In the first two cases—the achievement of thermal equilibrium by water and copper, and the bending of an electrical conductor—the kind of causal explanation that would satisfy proponents of the ontic conception can be supplied readily.

The third example is the 'explanation' of the generalized Balmer formula for the hydrogen spectrum on the basis of Bohr's atomic theory. It is not as straightforward as the other two. One can easily show, as van Fraassen does, how the formula for the line frequencies can be derived from Bohr's assumptions about quantized electron orbits and transitions between them. Nevertheless, because of the ad hoc character of these assumptions, it is difficult to believe that any explanation has been offered. "Why," we want to inquire, "can electrons occupy only certain orbits, but not any intermediate ones?" I suspect—given van Fraassen's emphasis upon the importance of rejections of requests for explanation—that his response would be, "Don't ask." This response would, I believe, be inappropriate. An answer (which, to be sure, was not available for more than a decade after Bohr first enunciated the old quantum theory) was provided by Louis de Broglie. Given the fact, subsequently confirmed by massive amounts of experimental data, that particles such as electrons exhibit wavelike behavior, one can explain Bohr's stationary states in terms of the existence of standing waves. The mechanism is essentially similar to that which accounts for the discrete frequencies that occur in the plucked strings of musical instruments. I am not claiming that de Broglie's explanation of the Bohr orbits is entirely satisfactory, but it did appeal to a mechanism that still plays a fundamental role in microphysics. All material particles

---

scattering). This limitation is serious, for we have no independent reason to think that explanation in quantum mechanics is essentially different from elsewhere." Considerations brought forth in the present chapter seem to me to provide ample grounds for the suspicion that explanations of quantum phenomena may be radically different from explanations of macroscopic phenomena. All of the problematic microphysical cases seem to involve quantum mechanical 'reduction of the wave packet'—a phenomenon that has no counterpart in macrophysics, to the best of my knowledge.

exhibit the basic characteristic of waves—namely, interference—and the new quantum mechanics, especially in the form provided by Schroedinger, is often called "wave mechanics." If the erotetic approach allows us to stop short of finding the underlying mechanisms, then, it seems to me, it does not constitute an adequate conception of scientific explanation.

Is it possible to provide causal explanations of quantum mechanical phenomena? I do not know. Van Fraassen argues cogently, on the basis of Bell's inequality and relevant experimental results, that "there are well-attested phenomena which cannot be embedded in any common-cause model" (1982a, p. 35). It appears that causal explanations are possible only if the concept of causality itself is fundamentally revised. John Forge (1982, p. 228) seems to evade the issue by suggesting that "a causal process is one which is governed by scientific laws (theories)." On this account of causality, every quantum mechanical process becomes causal by definition simply because quantum mechanics is a scientific theory. Such problems as the EPR paradox automatically disappear. Without endorsing it, van Fraassen (1980, p. 124) calls attention to a similar device. For purposes of argument, we might adopt the following definition:

the causal net = whatever structure of relations science describes.

We could then leave "to those interested in causation as such the problem of describing that structure in abstract but illuminating ways, if they wish." Since quantum theory describes many relations, it provides us with a significant portion of "the causal net." Again, a definition would be used to remove such perplexities as those generated by the EPR paradox.

O. Costa de Beauregard (1977) has suggested that the EPR correlation can be explained in terms of a causal influence traveling backward in time from the measurement on one particle to the earlier situation in which the two particles were in direct interaction with one another, and by another causal influence propagated forward in time from the point of interaction to the subsequent state of the other particle. Before accepting any such causal explanation, we would need strong evidence to show that causal influences can, indeed, be propagated backward in time. Other attempts to explain the EPR correlations appeal to 'hidden variables,' but all such explanations seem to encounter serious difficulties. To the best of my knowledge, there is no adequate explanation of the EPR correlations. The reason may be that we simply do not yet know enough about the microcosm to be able to explain such puzzling phenomena.

It would be premature, I believe, to conclude that causal explanations of quantum phenomena are impossible in principle. This conviction is *not* motivated by the hope or faith that a satisfactory 'hidden variable theory' will sooner or later be found. Rather, it seems to me, the nature and role of causality in microphysics is a deep and difficult matter to sort out. A great deal has been said, in both the philosophical and the physical literature, about the difficulties of extending causality into microphysics. Most such discussions are based upon the Heisenberg-Schroedinger-Born non-relativistic quantum theory, which is by no means the deepest microphysical theory available today. Before coming to any firm conclusions about causality, we should give careful consideration to quantum electrodynamics, the theory that covers interactions between charged particles and photons. This is, in some respects, the most satisfactory physical theory yet to be devised. Richard Feynman, in a systematic series of articles (1949, 1949a, 1950)—further elaborated in (Feynman, 1961, 1962)—attempted to provide a space-time formulation of quantum electrodynamics.

This theory involves obvious departures from causality as conceived in the context of classical physics—for example, it admits pair production and pair annihilation of particles; but since mass is not conserved in relativity theory, such occurrences do not even count as causal anomalies in relativistic quantum theory. Because energy is not strictly conserved in quantum mechanics, we have to contend with flocks of virtual particles, but their scope is severely limited by the uncertainty relation $\Delta E \Delta t \geqslant \hbar/2$. In Feynman's formulation, electrons *can* be said to scatter backward in time from regions of electromagnetic interaction, but with pair production and annihilation, it is *not necessary* to adopt this way of talking. An overall causal picture seems to emerge. Causal interactions occur in small regions of space-time. If causal anomalies are involved in these interactions, they are localized by the Heisenberg uncertainty relation. None of these interactions involves action-at-a-distance; indeed, in quantum field theory, a condition of Lorentz-invariance, which precludes causal interactions between regions that sustain a spacelike separation from one another, is imposed. Outside of localized regions of interaction, particles (including photons) behave as fairly reasonable causal processes. This picture of microphysical processes and interactions is not altogether at odds with the macrophysical account, developed in previous chapters, of causal processes and interactions. A detailed up-to-date treatment of the Feynman approach—which applies to the strong, weak, and gravitational forces, as well as the electromagnetic force—can be found in (Scadron, 1979). The basic features of microphysical causality are treated, sometimes explicitly and sometimes implicitly, in depth.

Quantum electrodynamics is not without difficulties, but it is a far deeper theory of microphysical mechanisms than was the early quantum mechanics of the mid-1920s. More recently, quantum chromodynamics has come to the fore as a theory of the interactions of hadrons. It bears some resemblance to quantum electrodynamics, but it is not nearly as fully developed. On the basis of my severely limited understanding of both theories, I cannot see any indication that quantum chromodynamics will fail to fit the general picture of causal processes and interactions I have sketched in connection with quantum electrodynamics. Thus it does not seem unreasonable to think—in the *most* tentative way—of microphysical explanation in terms of causal processes and causal interactions. Causal explanation may turn out, after all (*pace* Russell), not to be "a relic of a bygone age."

The central problem for quantum mechanical explanation, as I see it, rests with certain conservation laws. In classical physics, we have, for example, the law of conservation of momentum. When two particles collide (or enter into any other kind of interaction), momentum conservation requires a momentum transfer which occurs locally. The fact that causal interactions are constrained by *local* conservation of momentum does not seem mysterious. In the EPR problem, as originally propounded by Einstein, Podolsky, and Rosen, we have a composite system of two particles that has a fixed quantity of linear momentum, but according to the quantum theory, neither individual particle possesses a precise amount of linear momentum. Nevertheless, when a measurement is made that fixes an exact value for the linear momentum of one particle, the system somehow manages to provide the other particle with precisely the amount of linear momentum needed to maintain the total value for the system as a whole. It is the *remote* conservation—in conjunction with the fact that the subsystems do not at all times have precise amounts of linear momentum—that is perplexing. By what mechanism, we feel compelled to ask, does nature contrive to insure the conservation of momentum in this remote case?

Consider another simple and familiar example. An electron impinges upon a potential barrier of a sort that allows the electron a 50-50 chance of being transmitted or reflected. The wave that describes this electron is spread out so that half of it is on one side of the barrier, while the other half is on the other side. If the electron is subsequently detected on the far side of the barrier—signifying that in this instance it was transmitted rather than reflected—then the portion of the wave corresponding to a reflected electron instantaneously vanishes. Again, we have a case of remote satisfaction of a conservation law—the law, in this instance, being conservation of lepton number. There is one and only one electron. If it is not annihilated by encountering a positron (in which case lepton number is conserved) and if we do not allow any additional particles to enter the picture, then the conservation of lepton number implies that the detection of an electron on one side of the barrier must result in the disappearance of that portion of the wave that is on the other side of the barrier. The 'collapse of the wave function' can thus be construed as another instance of the remote fulfillment of a conservation law.[6]

It is tempting to maintain, with respect to the two foregoing examples, that no such remote conservation of physical quantities is involved. In the first case, some people may want to claim that each subsystem of the composite system always has a precise amount of linear momentum, so the conservation of momentum amounts to no more than the statement that each subsystem has a certain amount of linear momentum that is conserved. In the second case, they may want to claim that the electron always has a definite position on one side of the potential barrier or the other, and that the so-called collapse of the wave function amounts to no more than our discovery of where the electron is located. The fundamental import of the EPR example, as I see it, is to undermine this approach as a general strategy for escaping remote fulfillment of conservation laws. As the EPR experiment was reformulated by Bohm, making use of the different components of spin, this approach is blocked. The electron cannot simultaneously have both an eigenvalue of $+1/2$ or $-1/2$ for the z component of spin *and also* an eigenvalue of $+1/2$ or $-1/2$ for the y component of spin. Hence the perfect negative correlation between the spin of the positron and the spin of the electron requires *remote* fulfillment of conservation of angular momentum. The basic philosophical question seems to be this: Is local conservation of a physical quantity any less mysterious than remote conservation; is remote conservation any more miraculous than local conservation?[7]

Suppose that you have a bank account in each of two different banks. Both of the banks seem to have sloppy bookkeeping methods. You keep careful records of your deposits and withdrawals. When the monthly state-

---

[6] In a sober evaluation of the significance of this issue, Stein (1982) remarks, "Because I consider the issue of wave-packet reduction to be both serious and entirely unresolved—in particular, because I think we have at present *no serious evidence one way or the other* on the question of whether there is truly in nature any such *process* as 'reduction'—I believe that speculation of a general kind on the philosophical implications of reduction should indeed be admitted into consideration, but on the other hand should be considered only with great caution and reserve'' (Italics in original).

[7] Remote conservation avoids a basic difficulty that arises in connection with action-at-a-distance. If we wish to claim that A instantaneously causes B at some distant place, we will face the problem—because of the relativity of simultaneity—of specifying just when B is supposed to occur. In applying remote conservation laws, these temporal problems do not arise. If the subsystems S_1 and S_2 together contain a given amount of some physical quantity—say, angular momentum—then that amount must show up in measurements on the two subsystems whenever they are conducted. It does not matter whether the one measurement is made earlier than, later than, or simultaneous with the other. Remote conservation will not yield a method for sending information from one measurement to the other, and, hence, will not conflict with special relativity.

ments arrive from one of the banks, the reported balances fluctuate in an apparently random fashion about (what you consider) the true balances. The same thing happens if you make a balance inquiry at the banking machine between the monthy statements. The other bank behaves in the same way. Both the monthly statements and the interim balance inquiries exhibit the same sort of random fluctuation. Strangely, however, on any given day, the sum of the two balances quoted by the banks is precisely equal to the correct sum. Astonished by this remarkable 'conservation phenomenon,' you investigate carefully. At first, you suspect that the two banks communicate with one another very frequently, comparing notes to make sure that one bank compensates for the other's errors. Investigation shows, however, that no such communication occurs; there is no direct causal connection. Next, you suspect that although the fluctuations appear to be random, they are actually made in a systematic fashion, so that precisely compensating errors are 'programmed into' the two separate accounting systems. Careful investigation reveals that the errors at each bank are generated by some sort of random device—operating, for example, in response to a radioactive decay source. If there were a common cause, the random device would screen off the errors from the common cause. To say the least, such 'remote conservation' would be deeply perplexing.

Microphysical systems seem to operate in just such ways. A given system contains a certain amount of some physical quantity—say linear momentum. If a particular amount turns up at one place in the system, that constrains the amount that can appear elsewhere. Does this kind of conservation require explanation by means of some special mechanism? *Or is this one of the fundamental mechanisms by which nature operates?* These questions strike me as profound, and I make no pretense of having an answer to them.

Whether such principles as spatiotemporal continuity of causal processes are ultimately tenable depends upon the future of physical theory, a topic on which I am not qualified to speculate. The situation regarding action-at-a-distance is not altogether clear. Perhaps the very idea of a four-dimensional space-time is legitimate only for the macrocosm. Steven Weinberg, who received the Nobel Prize in 1979 for his work on unified field theories—and who *is* entitled, if anyone is, to speculate on such questions—recently offered some pertinent considerations. Speaking of a future theory that would embrace the strong, weak, electromagnetic, and gravitational forces, he remarked, ''I think it is reasonable to suppose that . . . the ultimate physics which describes nature on smaller scales will in fact be of a geometric nature.'' It may turn out, however, that the space described by this geometry will not be of the familiar variety. ''. . . I strongly suspect," he continues, "that ultimately we will find that the four-dimensional nature of space-time is another one of the illusory concepts that have their origin in the nature of human evolution, but that must be relinquished as our knowledge increases" (Weinberg, 1979, p. 46). Or, as Howard Stein recently declared (1982), "I am tempted to expect that we may yet have surprising things to learn about the concept of space-time, and about the character of what we are accustomed to call 'causation.' "

EXPLANATION AND UNDERSTANDING

Perhaps the most important fruit of modern science is the understanding it provides of the world in which we live, and of the phenomena that transpire within it. Such understanding results from our ability to fashion scientific explanations. In the first chapter, I raised a fundamental philosophical query about the nature of scientific explanation, namely, what sort of knowledge is explanatory knowledge, and on what basis can we say that it constitutes or contributes to our understanding? This same question—which has received surprisingly little explicit attention in the literature—was posed directly in a seminal article by Michael Friedman: "What is it about . . . scientific explanations . . . that gives us understanding of the world—what is it for a phenomenon to be scientifically understandable?" (1974, p. 5). He offers an appealing answer that has been in the background of many discussions, but that previously had not been systematically developed to any significant extent—namely, that unification of many diverse phenomena under a small number of generalizations is the key. "I claim that this is the crucial property of scientific theories we are looking for; this is the essence of scientific explanation— science increases our understanding of the world by reducing the total number of independent phenomena that we have to accept as ultimate or given. A world with fewer independent phenomena is, other things equal, more comprehensible than one with more" (1974, p. 15). Although Friedman's technical account of unification proved faulty (see Kitcher, 1976), Philip Kitcher (1981) has recently offered another attempt to implement this approach in terms of a different account of unification. Since the proposals of Friedman and Kitcher bear certain basic resemblances to the theories of Hempel and van Fraassen, their efforts must be associated with the epistemic conception of scientific explanation; indeed, their general approach seems to offer the best hope for an answer to the fundamental question that is available to those who adhere to any version of the epistemic conception.

The most severe shortcoming of the unification thesis—at least as it comes out when it is associated with the epistemic conception—is that it makes no reference to the physical mechanisms responsible for the phenomena that are to be explained. According to Friedman (1974, pp. 14–15), the explanatory import of the molecular-kinetic theory of gases lies, not in the fact that it informs us about the underlying mechanisms in their behavior, but, rather, in the fact that it unifies a number of gas phenomena under one generalization, and it connects the behavior of gases to the behavior of other bodies that obey Newtonian dynamical regularities.

According to the ontic conception, phenomena are explained by showing how they fit into the regularities in nature. Now if subsumption under regularities were the whole story, the foregoing account of explanatory power in terms of unification under a small number of broad regularities would apply to this approach as well. As we have seen, however, subsumption is not the whole story. It is not the whole story because, in the first place, there are pseudo-processes as well as causal processes. Pseudo-processes *are* regularities, but they need to be explained, for they lack explanatory import on their own. They do not transmit their own regularities; they are parasitic upon other causal processes for their very existence. Causal processes, in contrast, play a basic role in our account of explanation. In the second place, subsumption is not the whole story because there are noncausal lawlike regularities—such as the regular relation between airplane takeoff distance and the speed of drying of clothes hung out on a line, or the regular relation between the behavior of the tides and the position and phase of the moon—that need to be explained in terms of underlying causal mechanisms. Until these mechanisms are understood, such uniformities do not explain anything. In the third place, subsumption is not the whole story because there are statistical relevance relations that do not arise directly out of causal relations—recall Suppes's spurious probabilistic causes (chap. 7)—as well as statistical regularities that do directly mirror causal relations. The moral to be drawn is that underlying causal mechanisms hold the key to our understanding of the world. If subsumption under extremely general uniformities were the whole story, then the problem of local versus remote conservation laws would pose no explanatory puzzle at all. The perplexing features of quantum phenomena are fully subsumed under the laws of quantum mechanics, but as we have seen, such subsumption does not *explain* what goes on at the microphysical level.

As I said in chapter 2, scientific explanation appears to be a two-tiered affair. If an explanation is sought to account for some particular occurrence, the first step is to seek out prior events or conditions that are statistically relevant to the explanandum-event. If we can satisfy the desideratum of incorporating that event into an objectively homogeneous reference class—after the fashion discussed at length in chapters 2 and 3—then we have a statistical-relevance basis from which to work. It should be recalled that in constructing the S-R basis, all screening-off relations must be taken into account; consequently, factors rendered statistically irrelevant will not show up in the basis.

If we have correctly identified those factors that are statistically relevant to the event-to-be-explained, we have completed the bottom tier of our explanatory structure. A completed S-R basis—one that incorporates all statistically relevant factors—is, of course, something we rarely possess; it is an idealization. Nevertheless, as (Koertge, 1975) argues persuasively, it is philosophically important to have an adequate account of the ideal statistical basis.

The next step is to provide causal accounts of the statistical relevance relations involved in the S-R basis. There is no presumption that the causal relations can be 'read off' in any automatic or routine fashion. We must formulate causal hypotheses and apply standard scientific procedures to test them. According to the causal/mechanical approach I have been advocating, there are fundamental sorts of causal mechanisms that figure crucially in providing scientific explanations of statistical relevance relations. Given a statistical correlation between events of type A and events of type B, there may be a direct causal connection. Causal connections are causal processes. In the case of a direct causal connection, there must be a causal process connecting any given event A_i with the corresponding event B_i; the causal influence may be transmitted from A_i to B_i or it may go in the opposite direction. The fundamental criterion of capacity for mark transmission (as presented in chapter 5) must be applicable, at least in principle, to show that the process in question is, indeed, a causal process. Causal processes transmit energy, information, structure, and causal influence; they also transmit propensities to enter into various kinds of interactions under appropriate circumstances. A moving billiard ball carries a propensity to interact with another billiard ball in specifiable ways if their paths intersect, a golf ball carries a propensity to be deflected in specifiable ways if it encounters a tree branch, and an energetic photon carries a propensity to interact with an electron in a Compton scattering event. Some of these propensities may be deterministic; others seem clearly to be probabilistic.

If the statistical relevance relation between A and B that we are trying to explain does not arise out of a direct causal relation, we search for an indirect causal relation that obtains on account of a common cause. Causal forks of the conjunctive and interactive varieties (as discussed in chapter 6) constitute the mechanisms. If an interactive fork is invoked, there must be an intersection C of processes that lead from C to A and B, respectively. C must, of course, be a causal interaction, and the processes must be causal processes. Although the criterion CI (chapter 6) was formulated in terms of x-type intersections, other configurations of intersections may occur.

Causal processes and causal interactions are governed by basic laws of nature. Photons, for instance, travel in null geodesic paths unless they are scattered or absorbed upon encountering material particles. Freely falling material particles follow paths that are nonnull geodesics. Linear and angular momentum are conserved when particles interact with one another. Energy is conserved in isolated physical systems. *The causal/mechanical version of the ontic conception of scientific explanation is as much a covering-law conception as is any version of the epistemic conception.*

Interactive common causes explain certain kinds of statistical relevance relations; conjunctive common causes account for other kinds of statistical correlations. As I explained in some detail in chapter 6, conjunctive forks characterize regularities that occur because of independent causal processes that arise out of special background conditions. Because of the close relationship between conjunctive forks and the second law of thermodynamics, we have strong reason to believe that conjunctive forks have considerable explanatory force with respect to the order that exists in the universe. As I also explained in chapter 6, it appears that conjunctive forks cannot be characterized adequately in terms of statistical relations alone. In conjunctive forks, as in interactive forks, it is essential that there be causal processes that connect the common cause with the correlated effects.

It is important to acknowledge explicitly that considerations of physical detail—often quantitative in character—must be brought to bear in concrete situations. The mere existence of a causal process connecting two events does not necessarily explain anything; it must be a process of an appropriate sort. The fact that electromagnetic waves from the radio transmitter of a passing supersonic jet airplane impinge upon the plaster wall of a house does not explain the cracking of the plaster; the fact that a sonic boom from the same airplane interacts with the wall may constitute a perfectly adequate explanation of that crack.

Causal processes and causal interactions are, in many cases at least, probabilistic in character. As Rutherford discovered, an alpha particle impinging upon a thin gold foil has a certain probability of being deflected at a large angle. Subsequent discoveries in nuclear physics have only reinforced the supposition that interactions of this sort are ineluctably statistical. A carbon 14 atom is a causal process; in any fixed period of time, it has a definite probability of undergoing spontaneous decay. As I emphasized in chapter 7, causal processes transmit probability distributions of various sorts, in particular, probability distributions to undergo various kinds of modifications spontaneously or as a result of the sorts of interactions into which they may enter. When precise numerical values are available, they are, of course, furnished by detailed scientific theories.

Just as we have admitted that a complete S-R basis is something we can seldom (if ever) provide, so also must we grant that a complete causal account of those statistical relevance relations can seldom (if ever) be furnished. If we were to take the attitude that we never possess a genuine scientific explanation of any phenomenon unless we have a complete S-R basis as well as a complete causal account of the relevance relations in that basis, we would face the possibility—indeed, the likelihood—that our characterization is so stringent as to lead to the consequence that science never has provided any actual explanations, and probably never will be able to do so. A philosophical theory of scientific explanation that led to that consequence would have a serious mark against it.

There are, I believe, two recent suggestions that will be especially helpful in coming to terms with this problem. The first of these is Railton's concept of an *ideal explanatory text* (1981, p. 240). Although there are some matters of detail on which I would differ with Railton's account, there is a substantial amount of agreement—in particular, he advocates an ontic/mechanistic conception of explanation quite similar to the one I have been elaborating. "If one inspects the best-developed explanations in physics or chemistry textbooks and monographs," he remarks, "one will observe that these accounts typically include not only derivations of lower-level laws and generalizations from higher-level theory and facts but also attempts to *elucidate the mechanisms* at work" (1981, p. 242, italics in original). Speaking of his own deductive-nomological-probabilistic model of explanation, he says:

> The general form of the ideal D-N-P text is meant to represent the ideal striven for in actual explanatory practice, i.e., it comprises the things that a research program seeks to discover in developing the capacity to produce better explanations of chance phenomena. Thus, the ideal D-N-P text reflects not only an ideal of explanation, but of *scientific understanding*: we may say that we understand why a given chance phenomenon occurred to the extent that we are able, at least in principle, to produce the relevant ideal D-N-P text or texts. (1981, pp. 243–244, italics in original)

Precisely parallel considerations show how Hempel's basic D-N model can be conceived as a schema for an ideal explanatory text of nonprobabilistic phenomena (Railton, 1981, pp. 246–247).

[Is it] preposterous to suggest that any such ideal could exist for scientific explanation and understanding? Has anyone ever attempted or even wanted to construct an ideal causal or probabilistic text? It is not preposterous if we recognize that the actual ideal is not to *produce* such texts, but to have the ability (in principle) to produce arbitrary parts of them. It is thus irrelevant whether individual scientists ever set out to fill in ideal texts as wholes, since within the division of labor among scientists it is possible to find someone (or, more precisely, some group) interested in developing the ability to fill in virtually any particular aspect of ideal texts—macro or micro, fundamental or "phenomenological," stretching over experimental or historical or geological or cosmological time. (P. 247, italics in original)

We may agree with Railton that the notion of an ideal explanatory text is sensible and important, and that it does not embody pragmatic factors. The account of scientific explanation I have been attempting to elaborate corresponds to Railton's ideal explanatory text. This ideal is analogous to the ideal of rigorous mathematical demonstration. Mathematical 'proofs' are hardly ever strictly rigorous, but they should provide enough steps to enable the competent reader, at least in principle, to fill the gaps. Even though mathematicians seldom construct strictly rigorous demonstrations, the modern logical concept of proof has provided us with a considerably deeper understanding of mathematical proof than had previously been available. Similarly, a clear conception of the nature of the explanatory ideal should deepen our grasp of the notions of scientific explanation and scientific understanding.

Explanatory information is, in Railton's terminology, the kind of scientific knowledge that could play a part in constructing an ideal explanatory text. Most actual requests for scientific explanation—most explanation-seeking why-questions—are requests, not for ideal explanatory texts, but, rather, for explanatory information. We must admit, in agreement with van Fraassen, that pragmatic or contextual factors play a large role in determining what sort of explanatory information is pertinent in any particular situation (cf. Railton, 1981, pp. 245–247). The term "scientific explanation" itself is ambiguous; sometimes it refers to some appropriate explanatory information supplied in a given context, but sometimes it refers to an ideal explanatory text. Under the first construction, there are such things as genuine scientific explanations; under the second, it may be that genuine scientific explanations represent ideals that may sometimes be approached but are (almost) never fully realized. Railton's discussion contributes importantly to the clarification of the relationships among these concepts.

A second valuable suggestion has been offered by Paul Humphreys (1981, 1983) and Ben Rogers (1981); it concerns the relationship between precise numerical values of probabilities or degrees of statistical relevance on the one hand and causal relations on the other. Rogers' basic point can be summarized by saying that while precise values of probabilities are indispensable aids in the discovery and testing of causal hypotheses, only the causal relations have genuine explanatory import. I suspect that Rogers is entirely correct. Applied to the theory of scientific explanation I have been trying to elaborate, this thesis would imply that the S-R basis provides fundamental evidence regarding causal explanations, but the S-R basis itself does not constitute an ingredient in scientific explanations. When we consider the brutal complexity of the set of all statistical relevance relations that would obtain in almost any actual case, the idea of excluding the complete S-R basis from the explanation in question has almost irresistible appeal.

Relying upon considerations closely related to the suggestion of Rogers, Paul Humphreys has recently begun articulation of a theory of *aleatory explanation*. A basic feature of Humphreys' theory is that a satisfactory causal explanation of statistical occurrences requires identification of probabilistic causes that are positively relevant (contributing causes) and probabilistic causes that are negatively relevant (counteracting causes). Using these concepts, Humphreys suggests that "the canonical explanatory form for probabilistic explanations is '*A* because ϕ, despite ψ,' where ϕ is a nonempty set of contributing causes, ψ a set, possibly empty, of counteracting causes, and *A* is a sentence describing what is to be explained" (1981, p. 227).[8] Numerical values of probabilities and degrees of relevance are not required in explanations of this sort. Consider a somewhat modified version of an example Humphreys uses. The car went off the road because it was going fast and the road surface was snow-covered, despite the fact that the driver was skillful and the car was equipped with snow tires. We may reasonably assume that the precise degree of relevance of the snow tires to this situation would vary somewhat, depending upon the kind of car involved, but that their presence is nevertheless negatively relevant to the occurrence of the skid in any of these cases. There are, no doubt, many different snow conditions—soft or packed, wet or dry, shallow or deep— that would make a difference to the precise degree of relevance of the snow on the road, but it is positively relevant to the skid regardless. As Humphreys remarks, "We can be in ignorance of, and hence fail to specify in an explanation, any number of factors which raise or lower the probability, just so long as they do not defeat *X* as a contributing cause—or

[8] I consider Humphreys' theory, which accords explanatory force to negatively relevant as well as positively relevant causes, far more satisfactory than that of Tuomela (1981, p. 277), which holds that negatively relevant factors do not have explanatory value.

analogously, Y as a counteracting cause—of A in the circumstances'' (1981, p. 230).

Let us reconsider, in the light of Humphreys' approach, our example of juvenile delinquency that was used in chapter 2 to illustrate the idea of the S-R basis. Suppose we have assembled all of the statistical relevance relations of which we are aware—we have the explanans-partition with the associated probability values relating such factors as marital status of parents, religious background, type of residential community, socioeconomic status, and so on, to the occurrence of delinquency. Suppose we find, in examining the probability values associated with all of the cells in our homogeneous partition, that having parents who were never married is positively relevant under all other circumstances to delinquency. Then, according to Humphreys, we do not need to give all of the different probabilities for all of the different cells to cite the fact that his parents were never married as a contributing cause of Albert's delinquency. I am assuming, of course, as was stipulated in constructing the S-R basis in the first place, that this fact about the marital status of his parents is not screened off by other factors. The S-R basis contains an enormous amount of detailed numerical probability information about the bearing of many factors upon delinquency, but it would be otiose, according to Humphreys, to include it in the causal explanation as long as we are careful not to omit factors that can make the difference between a given circumstance being a contributing cause, a counteracting cause, or a factor that is causally irrelevant. Humphreys has summarized his view in the following terms: ''The conclusion I wish to draw is that probabilities are a means and not an end to many explanations. . . . I do believe that it is time to recognize probability for what it is in explanatory contexts—a widely used tool, rather like Fourier series or differential equations, and that it has no more of a central explanatory role than do those techniques'' (Humphreys, 1983, final paragraph).

The theory of aleatory explanation being developed by Humphreys is, obviously, a theory of probabilistic causal explanation, and I am strongly inclined to think that it is following the right track. What this means, I believe, is that a given ideal explanatory text—to adopt Railton's terminology—need not, and usually does not, include the corresponding complete S-R basis as a part. We can still conceive of the ideal text of an aleatory explanation that incorporates all of the contributing and counteracting causes, though this ideal may seldom or never be actually realized. If Humphreys' theory has a serious shortcoming, it would be, in my opinion, that he places insufficient emphasis upon the actual mechanisms—such as causal processes and causal interactions—that are needed, in addition to relations of positive and negative statistical relevance, to deal adequately with probabilistic causality. This inadequacy, if such it is, can be remedied by supplementation; no major excision seems to be required. Humphreys' aleatory theory could, for example, be combined advantageously with Railton's mechanistic approach.

THE CAUSAL/MECHANICAL MODEL

Throughout the foregoing discussions, I have tried to illustrate the main issues by means of a variety of examples—most of which have either historical or contemporary scientific importance—taken from many different disciplines. The selection of examples was motivated in part by a desire to exhibit something of the scope of the theory of scientific explanation that was being elaborated. In order to provide a concise summary of that theory, and to articulate the causal/mechanical model that emerges from it, I shall offer a set of examples including representatives from the behavioral sciences, the physical sciences, and the biomedical sciences. This will, I hope, enable those interested in such areas of science to see how the theory can be applied. The examples will also illustrate explanations both of particular facts and general regularities.

(1) As an example of explanation in the behavioral sciences, consider the case of a piece of bone found in Alaska that, on the basis of radiocarbon dating, seems to be about thirty thousand years old. It also shows clear signs of having been worked by a human artisan (Dumond, 1980). Let us assume, for purposes of this discussion, that the radiocarbon date is approximately correct, and that the bone is a human artifact. This object presents a problem, for none of the best authenticated sites of human habitation in the new world is more than twelve thousand years old, and even the more controversial age of the Meadowcroft Rockshelter is less than twenty thousand years (Fryman, 1982). Three potential explanations have some degree of plausibility. First, it may be that there has been human habitation in Alaska for at least thirty thousand years, and this piece of worked bone is the first evidence of such early habitation that has yet been found. Second, it is possible that there was a brief human incursion into Alaska about thirty thousand years ago, but that continuous human habitation began about twelve thousand years ago. Third, it is possible that although the bone is thirty thousand years old, it was frozen in an unworked condition for at least eighteen thousand years, and that it was discovered and worked by a human artisan no more than twelve thousand years ago.

I do not know which, if any, of the foregoing explanations is correct, and neither does anyone else, to the best of my knowledge. The point is simply that whichever explanation one chooses, it involves causal interactions and causal processes. In the case of the third explanation—which

I am inclined to prefer—we have the death of a large animal (e.g., a mammoth) and the deposition of the carcass in ice. There the bone remained undisturbed for many thousands of years. Later, the bone was discovered by a human artisan, who interacted with it. Subsequently, it was deposited in the site where it remained for several more millennia. This explanation is supported by the fact that contemporary inhabitants of Arctic regions use frozen bones of great age in making implements.

This explanation of a particular fact obviously involves the exhibition of causal mechanisms that led to the presence of the worked bone in the archaeological site. However, even if it is incorrect, the correct explanation will involve a different set of causal mechanisms. We should note explicitly that all of these putative causal explanations are probabilistic. In the third explanation, for example, it may be a matter of chance that the ancient artisan came across a piece of frozen bone suitable for working, and it is surely a matter of chance that this item escaped destruction in the period between its creation several thousand years ago and its recent discovery by archaeologists (Salmon, 1982c).

This explanation appeals to many objects, processes, and events that are unobserved by us—for example, human beings, an animal, the death of the animal, ice, the discovery of the bone, the fashioning of the bone—but some of them were observed by other people and all of them are observable in principle. The fact that a radiocarbon dating technique was used to establish the age of the bone does not mean that we are involved in any microphysical explanation. Microphysics is invoked to ascertain the age of the bone, but not to explain its presence in the site where it was discovered.

(2) In chapter 1, as an example of a regularity that needs explanation but has no explanatory force until it is causally explained, I mentioned the fact that there is a strict positive correlation between the amount of time required for clothes hung out on a line to dry and the distance required to get an airplane off of the ground at a nearby airport. I take it as given that the higher the relative humidity, the longer clothes will take to dry. Thus the phenomenon that requires explanation is the fact that increased relative humidity tends to make for a greater takeoff distance, other things—such as the temperature, the altitude of the airport, the type of plane, and the load it is carrying—being equal.

In the case of a propeller-driven airplane, both the lift imparted by the wings and the thrust imparted by the propeller are manifestations of the Bernoulli principle. According to this principle, the greater the velocity of a moving fluid, the smaller is the pressure that it exerts in the direction perpendicular to its direction of flow. The magnitude of this effect varies with the density of the fluid; the denser the fluid, the greater will be the lift and thrust. Consequently, the denser the air, the more readily will an airplane become airborne. In order to provide the explanation we are seeking, we must therefore show why humid air is *less dense* than dry air. Avogadro's law, which is embedded in the kinetic-molecular theory of gases, will enable us to do the job.

According to Avogadro's law, for fixed values of pressure and temperature, a given volume of gas contains the same number of molecules, regardless of the type of gas it is. A cubic meter of oxygen contains the same number of molecules as a cubic meter of nitrogen; a cubic meter of dry air contains the same number as a cubic meter of moist air. The main difference between moist air and dry air is that moist air contains more molecules of water and fewer nitrogen and oxygen molecules. The molecular weight of nitrogen (N_2) is 28, that of oxygen (O_2) is 32, and that of water (H_2O) is 18. When oxygen or nitrogen molecules are replaced with water molecules in a given volume of air, the mass is decreased; consequently, the density is lessened and the efficiency of the airfoils is reduced. This explains why a greater takeoff distance is needed when the humidity is higher.

The causal character of this explanation lies within the molecular-kinetic theory of gases. According to this theory, any gas is composed of particles in rapid motion that behave according to Newton's laws. Each molecule is a causal process that participates in causal interactions when it collides with another molecule or with a solid object such as the wall of its container or the wing of an airplane. These are the underlying causal mechanisms. The explanation obviously involves unobservable entities and it is fundamentally statistical, for it appeals to the average behavior of large numbers of such entities.

The foregoing examples illustrate two important aspects of causal explanation. The first example—the worked bone—involves the placing of the explanandum in a causal network consisting of relevant causal interactions that occurred previously and suitable causal processes that connect them to the fact-to-be-explained. Borrowing a term from Larry Wright (1976), I shall call explanations of this sort *etiological explanations*. Etiological explanations are, of course, thoroughly causal; they explain a given fact by showing how it came to be as a result of antecedent events, processes, and conditions. An important aspect of the theory of etiological explanations is the fact that—if Wright's analysis of functional explanations is satisfactory—functional explanations constitute a subset of etiological explanations.[9]

The second example—the relation between takeoff distance and drying

---

[9] It had been my original intention to offer an explicit treatment of functional explanation in this book, but I am so thoroughly persuaded of the fundamental correctness of Wright's treatment of the topic (1976), and of my inability to improve upon it, that I have decided against including it.

time—is an explanation involving several stages. We began by establishing the causal relationship between relative humidity and drying time for clothes, and the causal connection between the density of the air and the efficiency of airfoils. The crucial remaining part of the explanation depended upon the relationship between relative humidity and air density. In order to supply this part of the explanation, we noted the fact that air is a gas composed of molecules that behave in accordance with Newton's laws. These molecules are causal processes, and they participate in causal interactions. The explanation therefore rests upon the causal behavior of the constituents of air rather than upon a set of antecedent causes that are connected to the fact-to-be-explained.

Let us use the term *constitutive explanation* to refer to explanations of this sort. A constitutive explanation is thoroughly causal, but it does not explain particular facts or general regularities in terms of causal antecedents. The explanation shows, instead, that the fact-to-be-explained is constituted by underlying causal mechanisms. Many cases of physical reduction qualify as constitutive explanations. When, for example, we explain optical phenomena in terms of Maxwell's electromagnetic theory, the explanation is constitutive. Light waves *are* the electromagnetic waves (in a particular part of the spectrum) treated by Maxwell's theory. When, however, we explain the tides in terms of the gravitational influence of the moon and the sun, the explanation is etiological rather than constitutive, for the relative positions of the moon, the sun, and the earth are antecedent conditions that provide a causal explanation of the behavior of bodies of water on the surface of the earth.

The explanation of the presence of a worked bone that is thirty thousand years old in an Alaskan archaeological site can be considered a pure case of an etiological explanation. To explain this fact, it is not essential to look for causal constituents of the bone. The explanation of the fact that humid air is less dense than dry air can be considered a pure case of a constitutive explanation, for no causal antecedents had to be invoked in order to construct a satisfactory causal explanation of this phenomenon. An etiological explanation is an exhibition of the causal connections between the explanandum and prior occurrences; such an explanation fits the explanandum into an external pattern of causal relationships. A constitutive explanation consists of an exhibition of the internal causal structure of the explanandum; such an explanation exposes the causal mechanisms within the explanandum.

We may expect, in general, that a given explanation will have etiological aspects and constitutive aspects as well. To give a full explanation of the destruction of Hiroshima near the end of World War II, it would be necessary to refer to an atomic bomb, and to explain the explosion in terms of the assembly of a critical mass of U^{235}. Such an explanation would embody constitutive aspects. The explosion is explained in terms of a self-sustaining chain reaction, and this notion is causally explained in terms of the mechanisms of nuclear fission. The same explanation of the destruction of Hiroshima would include reference to the dropping of the bomb from an airplane and the detonation by implosion of a critical mass of fissionable material at a certain place above the city. These are etiological factors because they are antecedent events that contributed causally to the occurrence of the explanandum-event.

An explanation closely related to this one provides an excellent example of explanation in terms of conjunctive common causes. Medical studies of the survivors of the Hiroshima and Nagasaki bombings have established the fact that the incidence of leukemia was directly related to proximity to the nuclear explosion. The higher incidence is explained in terms of increased probability of leukemia among people exposed to excessive radiation. The common cause in each city is the field of intense radiation produced by the detonation of the bomb. The increased incidence of leukemia in those populations is not a chance clustering; it is explained by the elevated levels of radiation in those places. The causal processes probabilistically linking the nuclear explosions to the physiological damage that eventuates in leukemia are the photons and other particles traveling from the nuclear reactions to the bodies of the victims.

(3) As an example from the biomedical sciences, consider the explanation of the Great Plague, which struck England in 1665. The plague began in the spring and spread throughout the city until, by December of that year, London was deserted and one hundred thousand people had died. Everyone who was able left the cities and fled to the country.[10]

Although it had long been known that epidemics of plague were associated with unusually high death rates among rats—a statistical relevance relation of great importance—the occurrence of such epidemics was not understood until the end of the nineteenth century. In the latter part of the century, that disease became pandemic—touching every continent except Antarctica. In 1894, Alexander Yersin isolated and identified the bacterium (*Pasteurella pestis*) that causes the disease. It was then found that plague is endemic to rodents, especially rats, and that it is spread to different parts of the world by rats traveling on ships. The chief mechanism of transmission is a flea that bites infected rats and then bites other organisms,

[10] As a result of the Great Plague, Cambridge University was closed during 1665 and 1666. Newton, who was a student there at that time, returned to his birthplace at Woolsthorpe. He later recalled those years as his most productive period, writing, "I was at the prime of my age for invention and minded Mathematics and Philosophy more than at any time since." The year 1666 is known as Newton's *Annus Mirabilis*.

but there are other modes of infection as well. The infected rats were probably brought to England on trading ships.

In this case, again, we find that the explanation is causal, and the causal mechanisms involve continuous processes—the movements of the rats, the fleas, and the bacteria. It appeals to entities—the bacteria—that cannot be observed by means of unaided human senses, but that can be observed with the aid of a microscope. It does not invoke submicroscopic entities. Causal interactions are also involved. The fleas interact with the rats and the humans by biting both, and the bacteria that are transmitted by the bites interact with the infected body. This explanation also appeals to conjunctive forks. If we ask why so many people contract plague at about the same time, the principal answer involves the presence of infected rats and fleas. The fleas transmit the infection from rats to humans; for the most part, the disease is not communicated from one human directly to another. Many of the causal mechanisms are probabilistic. Not all residents of London contracted the disease; some people seem to have greater resistance than others. Not every flea that bites an infected organism actually transmits the disease to another organism, and not every organism that comes in contact with an infected flea becomes infected. Epidemiology is a science that regularly provides examples of causal explanations of the probabilistic sort.

I have presented this third example as an explanation of the occurrence of a particular epidemic, but it could easily be modified into a general explanation of such epidemics. In the explanation of the Great Plague of 1665, particular antecedent conditions were mentioned, but to explain such epidemics in general, we need to give a general characterization of the kind of antecedent conditions that lead to such occurrences. In explaining the particular epidemic, one would—in principle, at least—be attempting to exhibit particular causal processes. We would want to know what ship or ships from what particular locale brought the infected rats to London. Whether we are treating the explanation of a particular epidemic or dealing with such epidemics in general, these antecedent conditions will be etiological factors.

Since we are attempting to explain the epidemic, not the death of this or that particular victim, we need not try to fill in all of the particular connections involving particular rats, particular fleas, or particular bacteria, but we do want relevant statistical facts about interactions between rats and fleas, and between fleas and humans. The explanation of these statistical regularities consists in exhibiting them as the statistical result of an aggregate of causal processes and interactions. Such internal analysis of the explanandum—the epidemic—contributes to the constitutive aspects of the explanation.

The explanation of epidemics of plague in general consists in showing that certain effects have certain probabilities of occurrence whenever certain causal conditions obtain. Given the fact that plague is endemic in the rodent (primarily rat) population, and given the ubiquity of rats and other rodents, it is doubtful that plague will ever be wiped entirely from the face of the earth. Plague does not, however, pose serious threats of becoming epidemic among humans, for its spread can be controlled by appropriate use of insecticides and rodenticides on ships. Moreover, the disease can readily be diagnosed in humans, and it can be treated successfully with antibiotics. An understanding of the mechanisms of infection and transmission of plague has made it possible to eliminate the danger of such epidemics as the Great Plague of London and the Black Death of fourteenth-century Europe, which, it is plausibly argued, caused the deaths of as many as twenty-five million people (Meyer, 1965).

Let me now attempt to provide a reasonably concise characterization of causal/mechanical explanation. I shall speak of a 'causal/mechanical model' of scientific explanation, but since, on this model, explanations are not the sorts of things that can be explicated entirely in semantical terms, we must not expect a formally precise characterization. Inasmuch as I shall not be attempting to formulate exact sufficient conditions for acceptable scientific explanations, the model I shall advocate is not one of the sort condemned by Achinstein (1983, chap. 5).

An explanation consists of two parts, an explanans and an explanandum. The explanandum is some particular or general phenomenon—an event, a process, a regularity, a coincidence. I shall use the term ''fact'' in a comprehensive way to refer to such phenomena. Adopting this usage, we shall have to recognize such particular facts as the occurrence of an epidemic or the occurrence of a Compton scattering event, as well as such general facts as the Bernoulli principle or the higher density of dry air as compared to humid air. As I construe the term ''fact,'' facts are not linguistic entities—unless, of course, we are talking about language, and not just using language to talk about other sorts of facts. Facts are features of the physical world that exist objectively whether or not we endeavor to explain them.

It is important to note, in this connection, that particular facts do not necessarily embody all of the features of the phenomena which are involved. For example, archaeologists are attempting to explain the presence of a particular worked bone at a site in Alaska. The relevant features of this explanandum are the fact that the bone is thirty thousand years old, the fact that it was worked by a human artisan, and the fact that it had been deposited in an Alaskan site. Many other features are irrelevant to this sought-after explanation. The orientation of the bone with respect to the cardinal points of the compass at the time it was discovered, its precise size and shape (beyond the fact that it was worked), and the distance of the site from the nearest stream are all irrelevant. It is important to realize that we cannot aspire to explain particular phenomena in their full particularity. The same consideration obviously applies to such explananda as the Great Plague and the bombing of Hiroshima. In explanations of particular phenomena, the explanation-seeking why-question—suitably clarified and reformulated if necessary—should indicate those aspects of the phenomena for which an explanation is sought.

If we are attempting to explain some regularity, such as the fact that tails of comets are generally directed away from the sun, we are not concerned with many of the characteristics that differentiate one comet from another. It does not matter whether the comet is large or small; whether its orbit is parabolic, hyperbolic, or elliptical; whether its tail is composed of ionized molecules or dust particles; whether (given a closed orbit) it returns seldom or frequently. This aspect of general regularities is usually an explicit feature of the ways in which they are described. When we say that almost all tails of comets are pointed away from the sun, we are clearly omitting reference to more specific properties of comets or their tails.

As I prefer to construe these terms, the explanandum is a fact—it is the fact-to-be-explained. I also consider it appropriate to refer loosely to events, things, processes, and regularities as explananda, provided reasonable care is taken in so doing. When we want to refer to the statement that a fact obtains—or that an event occurs, a thing exists, a process transpires, a regularity holds—we may speak of the explanandum-statement.

To explain a fact is to furnish an explanans. We may think of the explanans as a set of explanatory facts; these general and particular facts account for the explanandum. Among the explanatory facts are particular events that constitute causes of the explanandum, causal processes that connect the causes to their effects, and causal regularities that govern the causal mechanisms involved in the explanans. The explanans is a complex of objective facts. In presenting the explanans, we use statements that detail the explanatory facts—we may refer to them as explanans-statements.

The term "explanation" may be construed in either of two ways. It may be taken as the combination of the explanans and the explanandum—the explanans having its place in the causal pattern or causal nexus in which it is objectively embedded. This term may also be used to refer to the combination of explanans-statements and explanandum-statement—the linguistic entity that is used to present the objective facts. Railton's term "explanatory text" is an entirely suitable expression to employ when we want to talk about the statements used to formulate an explanation.

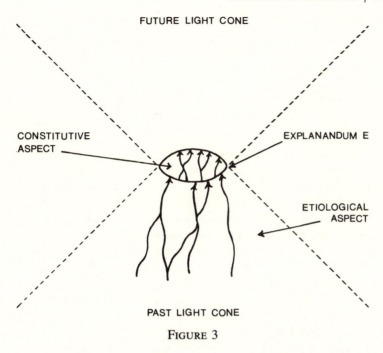

FIGURE 3

It may be helpful to think diagrammatically about causal/mechanical explanations. Suppose we want to explain some event E. We may look at E as occupying a finite volume of four-dimensional space-time (see Fig. 3). If we want to show why E occurred, we fill in the causally relevant processes and interactions that occupy the past light cone of E. This is the etiological aspect of our explanation; it exhibits E as embedded in its causal nexus. If we want to show why E manifests certain characteristics, we place inside the volume occupied by E the internal causal mechanisms that account for E's nature. This is the constitutive aspect of our explanation; it lays bare the causal structure of E. To explain a general regularity is much the same sort of operation. Instead of focusing upon one particular volume of space-time, we apply precisely the same considerations to any volume of space-time that is similar in relevant respects. The relevant similarities are given by the nature of the regularity we are trying to explain. As we have just seen, an explanation of a particular epidemic of plague is similar to an explanation of plague epidemics in general. Likewise, an explanation of the difficulty of getting a particular airplane off of the ground on a humid day closely resembles an explanation of the general relationship between the relative humidity of the air and the takeoff distance for aircraft.

Indeed, it seems altogether reasonable to maintain that any adequate explanation of a particular fact must be, in principle, generalizable into an explanation of a suitable sort of regularity.

THE FINAL CONTRAST

We can, I believe, make a fairly graphic comparison between the epistemic conception (in all of its various versions, but especially the inferential version) and the ontic conception construed in the causal/mechanical way. The epistemic conception looks upon the world as a place that exhibits various discoverable regularities, and when these regularities are known, the world can be seen as a dependable environment. We come to count on certain regularities, such as the sequence of seasons, and we can anticipate much that we will encounter as time elapses, as we go from place to place, and as we perform certain actions. This way of looking at the world leads naturally to an inferential view of scientific explanation, and to Hempel's celebrated 'symmetry thesis' of explanation and prediction. For purposes of inference, it is sufficient to be acquainted with the regularities that obtain in the world; it does not really matter what underlying mechanisms give rise to these regularities.

The ontic conception looks upon the world, to a large extent at least, as a black box whose workings we want to understand. Explanation involves laying bare the underlying mechanisms that connect the observable inputs to the observable outputs. We explain events by showing how they fit into the causal nexus. Since there seem to be a small number of fundamental causal mechanisms, and some extremely comprehensive laws that govern them, the ontic conception has as much right as the epistemic conception to take the unification of natural phenomena as a basic aspect of our comprehension of the world. *The unity lies in the pervasiveness of the underlying mechanisms* upon which we depend for explanation.

One of the major aims of this book has been to provide a 'clarification of the explicandum'—an attempt to exhibit in some detail what appear to be the fundamental and persistent conceptions of scientific explanation that have motivated much of the philosophical work on the subject, including the elaboration of various semiformal 'models of scientific explanation.' For reasons already sufficiently belabored, I have found the modal conception inadequate. That leaves us with the epistemic and the ontic conceptions. The epistemic conception has exercised an extraordinarily pervasive influence upon thought about scientific explanation for several decades at the very least. I have been urging adoption of the ontic conception. Right or wrong, this is no trivial matter; the shift from the epistemic to the ontic conception involves, in my opinion, nothing less than a radical gestalt switch.

The point can be illustrated quite simply. Take any of the many examples in which there is some irreducibly probabilistic mechanism that produces either of two outcomes, one with high probability and the other with low probability. The best example is still the old favorite, spontaneous radioactive decay. Suppose, then, that we have a tritium atom that was put into a closed box and left there for forty-nine years. Since its half-life is about 12¼ years, the probability that it will decay within that period is about 15/16, and the probability that it will survive intact is about 1/16. If, after forty-nine years, we open the box and find that the tritium atom has decayed, most philosophers are willing to say that we can explain that eventuality on the basis of its high probability, supplemented, perhaps, by a story about how radioactive decay takes place. If, however, the tritium atom has remained intact for the period of forty-nine years, the vast majority of philosophers, I venture to guess, will be intensely dissatisfied with a parallel answer, in terms of its low probability, to the question about why it did not decay. The difficulty seems to be that we cannot show "that it was to be expected," or why this particular atom failed to decay in that period of time given that the great majority of such atoms do. These feelings of uneasiness about the explanation of low-probability events arise, I am convinced, from a commitment (which may be quite unconscious) to the epistemic conception. If, however, we adopt the view that the point of an explanation is to exhibit events as having their places in patterns, and accept the notion that some of the patterns are irreducibly statistical patterns, then the explanation of the low-probability event seems just as natural and acceptable as that of the high-probability event, for the stochastic pattern is one that includes events of both types. If, in addition, our theories provide us with an account of the physical mechanisms that give rise to such statistical patterns, so much the better.

In chapter 4 I argued at length that the following attitudes toward scientific explanation, which arise naturally from the epistemic conception, are profoundly mistaken:

1. to insist, as Hempel did for many years, that an explanation of an occurrence must show that the fact-to-be-explained was to be expected;
2. to demand, with van Fraassen, that an explanation confer upon the fact-to-be-explained a favored position vis-à-vis various alternative eventualities;
3. to require that an explanation show why one outcome *rather than* an alternative occurred;

4. to reject the possibility that circumstances of a given type C can, on one occasion, explain the occurrence of an event of type E, and on another occasion explain the occurrence of an incompatible alternative outcome E'.

As I have tried to show, once the ontic conception is adopted, avoidance of the foregoing errors becomes easy and natural. If, as seems likely, we live in an indeterministic world, then our world demands a conception of scientific explanation that allows us to reject 1–4.

It is my hope that the causal theory of scientific explanation outlined in this book is reasonably adequate for the characterization of explanation in most scientific contexts—in the physical, biological, and social sciences—as long as we do not become involved in quantum mechanics. There are, no doubt, many points on which clarification or correction are required, but I am optimistic enough to think that the theory is adequate in its major outlines, and that the needed adjustments can be incorporated without undue stress on the overall structure. Time, and severe criticism, will tell.[11]

Since, as I pointed out at the beginning of this chapter, the advocate of the ontic conception is not committed to a search for 'the *logic* of explanation,' I have not been trying to lay down conditions that must be satisfied by all admissible scientific explanations in all possible worlds. Someone much cleverer than I might be able to achieve that feat, but it is surely beyond me. My aim has been to articulate contingent features of scientific explanations in this world as we presently conceive it. Nor have I tried to set forth criteria that must be satisfied by all admissible explanations in all domains of this world. It would be pleasant to think that the theory propounded in this book could be extended into the microcosm, but if that turns out to be impossible, it would not constitute an objection to its application within domains of science that do not involve microphysics.[12] It would come as no surprise to me to learn that a fully general philosophical theory of scientific explanation has to be constructed piecemeal, and that not all parts are available as yet. I would be gratified if the foregoing discussions furnished some of the pieces.

As the title of this chapter suggests, the philosophical framework within

[11] In his (1982), Glymour offers a number of constructive suggestions along this line; in my (1982b, pp. 272–275), I respond to them.

[12] Van Fraassen (1980, p. 123) kindly remarks, "The conclusion suggested by all of this is that the type of explanation characterized by Salmon, though apparently of central importance, is still at most a subspecies of explanations in general." I am gratified by this assessment. For reasons that have been spelled out in this chapter, I do not consider the fact that the enterprise has not yielded a fully general characterization of scientific explanation to be a serious criticism of what has been done. If this sort of partial success has been realized, I am genuinely pleased, for I could not have hoped for any more.

which I have been working can aptly be called—as Clark Glymour (1982) perceptively noted—"the mechanical philosophy." If we are to do justice to that honorable philosophical tradition in attempting to characterize scientific explanation in the context of our twentieth-century world picture, we must, of course, give due consideration to *the* great mechanical theory of the twentieth century—quantum mechanics. I have not offered any account of quantum mechanical explanation, and I do not believe that anyone else has done so either. The reason for this lacuna may be, very simply, that we still do not understand quantum mechanics well enough. In his address, "On the Present State of the Philosophy of Quantum Mechanics," delivered at the 1982 meeting of the Philosophy of Science Association, Howard Stein remarked, "The problem of reduction of the wave packet . . . remains what it always has been: baffling." If he is correct—and I feel fairly confident that he is—in claiming that we still do not understand so fundamental an aspect of quantum theory, it is hardly surprising that nothing approaching an adequate theory of quantum mechanical explanation has yet been offered. In any case, to provide a satisfactory treatment of microphysical explanation constitutes a premier challenge to contemporary philosophy of science—one that lies beyond the scope of this book and beyond my present capabilities.

Bibliography

This bibliography mentions only those materials that are cited in the text.

Achinstein, Peter
   1983   *The Nature of Explanation*. New York: Oxford University Press.
Aquinas, St. Thomas
   1947   *Summa Theologica*. New York: Benziger Brothers.
Aristotle
   1928   *Analytica Posteriora*. In W. D. Ross, ed., *The Works of Aristotle*, vol. 1. Oxford: Clarendon Press.
Arnauld, Antoine
   1964   *The Art of Thinking (Port-Royal Logic)*. Indianapolis, Ind.: Bobbs-Merrill.
Belnap, Nuel D., Jr., and Steel, J. B., Jr.
   1976   *The Logic of Questions and Answers*. New Haven, Conn.: Yale University Press.
Bergson, Henri
   1911   *Creative Evolution*. New York: Holt, Rinehart and Winston.
Bohm, David
   1951   *Quantum Theory*. Englewood Cliffs, N.J.: Prentice-Hall.
Bolton, Herbert E.
   1960   *The Rim of Christendom*. New York: Russell and Russell.
Braithwaite, R. B.
   1953   *Scientific Explanation*. Cambridge: At the University Press.
Bridgman, Percy W.
   1928   *The Logic of Modern Physics*. New York: Macmillan.
Brody, Baruch
   1975   "The Reduction of Teleological Sciences." *American Philosophical Quarterly* 12:69–76.
Bromberger, Sylvain
   1966   "Why-Questions." In Robert G. Colodny, ed., *Mind and Cosmos*, 86–111. Pittsburgh: University of Pittsburgh Press.
Cantrell, C. D., and Scully, Marlan O.
   1978   "The EPR Paradox Revisited." *Physics Reports* 43:500–508.
Carnap, Rudolf
   1950   *Logical Foundations of Probability*. Chicago: University of Chicago Press. 2nd ed., 1962.

1966   *Philosophical Foundations of Physics.* New York: Basic Books. Edited by Martin Gardner.

1974   *An Introduction to the Philosophy of Science.* New York: Basic Books. Reprint, with corrections, of (Carnap, 1966).

Cartwright, Nancy
   1983   *How the Laws of Physics Lie.* New York: Oxford University Press.

Chaffee, Frederic H., Jr.
   1980   "The Discovery of a Gravitational Lens." *Scientific American* 243, no. 5 (November):70–88.

Church, Alonzo
   1940   "On the Concept of a Random Sequence." *Bulletin of the American Mathematical Society* 46:130–135.

Coffa, J. Alberto
   1974   "Randomness and Knowledge." In Kenneth F. Schaffner and Robert S. Cohen, eds., *PSA 1972*, 103–115. Dordrecht: D. Reidel.

   1974a   "Hempel's Ambiguity." *Synthese* 28:141–164.

Cohen, L. J.
   1975   "Comment." In Stephan Körner, ed., *Explanation*, 152–159. Oxford: Basil Blackwell.

Cohen, S. N.
   1975   "The Manipulation of Genes." *Scientific American* 233, no. 1 (July):24–33.

Cooper, Leon N.
   1968   *An Introduction to the Meaning and Structure of Physics.* New York: Harper and Row.

Copeland, Arthur H.
   1928   "Admissible Numbers in the Theory of Probability." *American Journal of Mathematics* 50:535–552.

Copi, Irving M.
   1978   *Introduction to Logic.* 5th ed. New York: Macmillan.

Costa de Beauregard, O.
   1977   "Two Lectures on the Direction of Time." *Synthese* 35:129–154. Reprinted in (Salmon, 1979a).

Crow, James F.
   1979   "Genes that Violate Mendel's Rules." *Scientific American* 240, no. 2 (February):134–146.

Davies, P.C.W.
   1979   *The Forces of Nature.* Cambridge: At the University Press.

   1980   *The Search for Gravity Waves.* Cambridge: At the University Press.

d'Espagnat, Bernard
  1979   "The Quantum Theory and Reality." *Scientific American* 241 no. 5 (November):158–181.
Dumond, Don E.
  1980   "The archaeology of Alaska and the peopling of America." *Science* 209:984–991.
Einstein, Albert; Podolsky, B.; and Rosen, N.
  1935   "Can Quantum-Mechanical Description of Physical Reality Be Considered Complete?" *Physical Review* 47:777–780.
Emerson, Ralph Waldo
  1836   "Hymn Sung at the Completion of the Battle Monument, Concord."
Fair, David
  1979   "Causation and the Flow of Energy." *Erkenntnis* 14:219–250.
Fetzer, James H.
  1981   *Scientific Knowledge.* Dordrecht: D. Reidel.
  1981a "Probability and Explanation." *Synthese* 48:371–408.
Fetzer, James H., and Nute, D. E.
  1979   "Syntax, Semantics, and Ontology: A Probabilistic Causal Calculus." *Synthese* 40:453–495.
Feynman, Richard P.
  1949   "The Theory of Positrons." *Physical Review* 76:749–759. Reprinted in (Feynman, 1961) and (Schwinger, 1958).
  1949a "Space-Time Approach to Quantum Electrodynamics." *Physical Review* 76:769–789. Reprinted in (Feynman, 1961) and (Schwinger, 1958).
  1950   "Mathematical Formulation of the Quantum Theory of Electromagnetic Interaction." *Physical Review* 80: 440–457.
  1961   *Quantum Electrodynamics.* New York: W. A. Benjamin.
  1962   *The Theory of Fundamental Processes.* New York: W. A. Benjamin.
Feynman, Richard P.; Leighton, Robert B.; and Sands, Matthew
  1965   *The Feynman Lectures on Physics*, vol. 3. Reading, Mass.: Addison-Wesley.
Forge, John
  1982   "Physical Explanation: With Reference to the Theories of Scientific Explanation of Hempel and Salmon." In Robert McLaughlin, ed., *What? Where? When? Why?* 211–229. Dordrecht: D. Reidel.
Friedman, Michael
  1974   "Explanation and Scientific Understanding." *Journal of Philosophy* 71:5–19.

Fryman, R. F.
1982 "Prehistoric Settlement Patterns in the Cross Creek Drainage."
In R. C. Carlisle and J. M. Adovasio, eds., *Meadowcroft*, 53–
68. Pittsburgh, Pa.: Cultural Resource Management Program,
Department of Anthropology, University of Pittsburgh.

Gardner, Martin
1981 *Science: Good, Bad, and Bogus*. Buffalo, N.Y.: Prometheus
Books.

Gardner, Michael
1979 "Realism and Instrumentalism in 19th Century Atomism." *Philosophy of Science* 46:1–34.

Gehrenbeck, Richard K.
1978 "Electron Diffraction Fifty Years Ago." *Physics Today* 31, no.
1:34–41.

Giere, Ronald N.
1979 *Understanding Scientific Reasoning*. New York: Holt, Rinehart
and Winston.

Glymour, Clark
1980 *Theory and Evidence*. Princeton: Princeton University Press.
1982 "Causal Inference and Causal Explanation." In Robert Mc-
Laughlin, ed., *What? Where? When? Why?* 179–191. Dordrecht:
D. Reidel.

Good, I. J.
1961–1962 "A Causal Calculus (I & II)." *British Journal for the
Philosophy of Science* 11:305–318; 12:43–51. See also "Cor-
rigenda" 13:88.
1980 "Some Comments on Probabilistic Causality." *Pacific Philosophical Quarterly* 61:301–304.

Goodwin, Donald W., and Guze, Samuel B.
1979 *Psychiatric Diagnosis*. 2nd ed. Oxford: Oxford University Press.

Greeno, James G.
1970 "Evaluation of Statistical Hypotheses Using Information Transmitted." *Philosophy of Science* 37:279–293.
1971 "Theoretical Entities in Statistical Explanation." In Roger C.
Buck and Robert S. Cohen, eds., *PSA 1970*, 3–26. Dordrecht:
D. Reidel.
1971a "Explanation and Information." In (Salmon et al., 1971), 89–
104; reprint of (Greeno, 1970).

Grünbaum, Adolf
1973 *Philosophical Problems of Space and Time*. 2nd ed. Dordrecht:
D. Reidel.

Hacking, Ian
  1981 "Do We See Through a Microscope?" *Pacific Philosophical Quarterly* 62:305–322.
Halley, Edmund
  1720 "Of the infinity of the sphere of fix'd stars" and "Of the number, order, and light of fix'd stars." *Royal Society of London, Philosophical Transactions* 31:22–26.
Hanna, Joseph
  1969 "Explanation, Prediction, Description, and Information." *Synthese* 20:308–344.
  1978 "On Transmitted Information as a Measure of Explanatory Power." *Philosophy of Science* 45:531–562.
  1981 "Single Case Propensities and the Explanation of Particular Events." *Synthese* 48:409–436.
  1983 "Probabilistic Explanation and Probabilistic Causality." In Peter Asquith and Thomas Nickles, eds., *PSA 1982*, vol. 2. East Lansing, Mich.: Philosophy of Science Association.
Hanson, N. R.
  1958 *Patterns of Discovery*. Cambridge: At the University Press.
Harré, R., and Madden, E. H.
  1975 *Causal Powers*. Oxford: Basil Blackwell.
Hempel, Carl G.
  1959 "The Logic of Functional Analysis." In L. Gross, ed., *Symposium on Sociological Theory*, 271–307. New York: Harper and Row. Reprinted, with revisions, in (Hempel, 1965).
  1960 "Inductive Inconsistencies." *Synthese* 12: 439–469. Reprinted, with slight changes, in (Hempel, 1965).
  1962 "Deductive-Nomological vs. Statistical Explanation." In Herbert Feigl and Grover Maxwell, eds., *Minnesota Studies in the Philosophy of Science*, 3:98–169. Minneapolis: University of Minnesota Press.
  1962a "Explanation in Science and in History." In Robert G. Colodny, ed., *Frontiers in Science and Philosophy*, 7–34. Pittsburgh: University of Pittsburgh Press.
  1965 *Aspects of Scientific Explanation and Other Essays in the Philosophy of Science*. New York: Free Press.
  1965a "Aspects of Scientific Explanation." In (Hempel, 1965), 331–496.
  1966 *Philosophy of Natural Science*. Englewood Cliffs, N.J.: Prentice-Hall.
  1968 "Maximal Specificity and Lawlikeness in Probabilistic Explanation." *Philosophy of Science* 35:116–133.

1977 "Nachwort 1976: Neuere Ideen zu den Problemen der statistischen Erklärung." In Carl G. Hempel, *Aspekte wissenschaftlicher Erklärung*, 98–123. Berlin/New York: Walter de Gruyter.

Hempel, Carl G., and Oppenheim, Paul

1948 "Studies in the Logic of Explanation." *Philosophy of Science* 15:135–175; reprinted, with added Postscript, in (Hempel, 1965).

Hesslow, Germund

1976 "Two Notes on the Probabilistic Approach to Causality." *Philosophy of Science* 43:290–292.

Hilts, Victor

1973 "Statistics and Social Science." In Ronald N. Giere and Richard S. Westfall, eds., *Foundations of Scientific Method: The Nineteenth Century*, 206–233. Bloomington, Ind.: Indiana University Press.

Holton, Gerald, and Brush, Stephen G.

1973 *Introduction to Concepts and Theories in Physical Science*. 2nd ed. Reading, Mass.: Addison-Wesley.

Hume, David

1888 *A Treatise of Human Nature*. Oxford: Clarendon Press.

1955 *An Inquiry Concerning Human Understanding*. Indianapolis, Ind.: Bobbs-Merrill. Also contains "An Abstract of *A Treatise of Human Nature*."

Humphreys, Paul

1981 "Aleatory Explanations." *Synthese* 48:225–232.

1983 "Aleatory Explanations Expanded." In Peter Asquith and Thomas Nickles, eds., *PSA 1982*. East Lansing, Mich.: Philosophy of Science Association.

Jeffrey, Richard C.

1969 "Statistical Explanation vs. Statistical Inference." In Nicholas Rescher, ed., *Essays in Honor of Carl G. Hempel*. Dordrecht: D. Reidel. Reprinted in (Salmon et al., 1971), 19–28.

1971 "Remarks on Explanatory Power." In Roger C. Buck and Robert S. Cohen, eds., *PSA 1970*, 40–46. Dordrecht: D. Reidel.

Kelvin, (Lord), William Thomson

1884 *Notes of Lectures on Molecular Dynamics and the Wave Theory of Light*. Baltimore: Johns Hopkins University.

King, John L.

1976 "Statistical Relevance and Explanatory Classification." *Philosophical Studies* 30:313–321.

Kitcher, Philip

1976 "Explanation, Conjunction, and Unification." *Journal of Philosophy* 73:207–212.

1981 "Explanatory Unification." *Philosophy of Science* 48:507–531.

Koertge, Noretta

1975 "An Exploration of Salmon's S-R Model of Explanation." *Philosophy of Science* 42:270–274.

Körner, Stephan

1975 *Explanation*. Oxford: Basil Blackwell.

Kyburg, Henry E., Jr.

1961 *Probability and the Logic of Rational Belief*. Middletown, Conn.: Wesleyan University Press.

1965 "Comment." *Philosophy of Science* 32:147–151.

1970 "Conjunctivitis." In Marshall Swain, ed., *Induction, Acceptance, and Rational Belief*, 55–82. Dordrecht: D. Reidel.

Lambert, Karel

1980 "Explanation and Understanding: An Open Question?" In Risto Hilpinin, ed., *Rationality in Science*, 29–34. Dordrecht: D. Reidel.

Laplace, P. S.

1951 *A Philosophical Essay on Probabilities*. New York: Dover Publications.

Lecomte du Nouy, Pierre

1947 *Human Destiny*. New York: Longmans Green.

Lehman, Hugh

1972 "Statistical Explanation." *Philosophy of Science* 39:500–506.

Leibniz, G. W.

1951 "The Theodicy: Abridgement of the Argument Reduced to Syllogistic Form." In Philip P. Wiener, ed., *Leibniz Selections*, 509–522. New York: Charles Scribner's Sons.

1952 *Theodicy*. New Haven: Yale University Press.

1965 "A Vindication of God's Justice Reconciled with His Other Perfections and All His Actions." In Paul Schrecker, ed., *Monadology and Other Philosophical Essays*, 114–147. Indianapolis, Ind.: Bobbs-Merrill.

Leighton, Robert B.

1959 *Principles of Modern Physics*. New York: McGraw-Hill.

Lucretius

1951 *On the Nature of the Universe*. Baltimore: Penguin Books.

MacCorquodale, Kenneth, and Meehl, Paul E.

1948 "On a Distinction Between Hypothetical Constructs and Intervening Variables." *Psychological Review* 55:95–107.

Mach, Ernst

1914 *The Analysis of Sensations*. Chicago and London: Open Court.

Mackie, J. L.
    1974   *The Cement of the Universe*. Oxford: Clarendon Press.
Martin-Löf, P.
    1966   "The Definition of Random Sequences." *Information and Control* 9:602–619.
    1969   "Literature on v. Mises' Kollectivs Revisited." *Theoria* 35:12–37.
McCalla, R. L.
    1976   "Letter to the Editor." *Scientific American* 234, no. 2 (February):8.
McLaughlin, Robert, ed.
    1982   *What? Where? When? Why? Essays on Induction, Space and Time, Explanation*. Dordrecht: D. Reidel.
Mellor, D. H.
    1976   "Probable Explanation." *Australasian Journal of Philosophy* 54:231–241.
Meyer, K. F.
    1965   *Encyclopaedia Britannica*, 14th ed., s.v. "Plague."
Misner, Charles W.; Thorne, Kip S.; and Wheeler, John A.
    1973   *Gravitation*. San Francisco: W. H. Freeman.
Nagel, Ernest
    1961   *The Structure of Science*. New York: Harcourt, Brace and World.
Newton, Isaac
    1947   *Sir Isaac Newton's Mathematical Principles of Natural Philosophy and His System of the World*. Berkeley: University of California Press.
Niiniluoto, Ilkka
    1981   "Statistical Explanation Reconsidered." *Synthese* 48:437–472.
Nye, Mary Jo
    1972   *Molecular Reality*. London: Macdonald.
Otte, Richard
    1981   "A Critique of Suppes' Theory of Probabilistic Causality." *Synthese* 48:167–189.
Pagels, Heinz R.
    1982   *The Cosmic Code*. New York: Bantam Books.
Pauling, Linus
    1970   *Vitamin C and the Common Cold*. San Francisco: W. H. Freeman.
Pearson, Karl
    1957   *The Grammar of Science*. 3rd ed. New York: Meridian Books.
Peirce, C. S.
    1932   *The Collected Papers of Charles Sanders Peirce*. Edited by Charles

Hartshorne and Paul Weiss. 6 vols. Cambridge, Mass.: Harvard University Press.

Perrin, Jean

1913 *Les Atomes*. Paris: Alcan.

1923 *Atoms*. Translated by D. L. Hammick. New York: Van Nostrand.

Phillips, James C.

1982 "The Physics of Glass." *Physics Today* 35, no. 2 (February):27–33.

Popper, Karl R.

1935 *Logik der Forschung*. Vienna: Springer.

1959 *The Logic of Scientific Discovery*. New York: Basic Books. Translation of (Popper, 1935).

1972 *Objective Knowledge*. Oxford: Clarendon Press.

Putnam, Hilary

1982 "Why There Isn't a Ready-Made World." *Synthese* 51:141–167.

1982a "Why Reason Can't Be Naturalized." *Synthese* 52:3–23.

Quine, Willard van Orman

1951 "Two Dogmas of Empiricism." *Philosophical Review* 60:20–43. Reprinted in (Quine, 1953).

1953 *From a Logical Point of View*. Cambridge, Mass.: Harvard University Press.

Railton, Peter

1978 "A Deductive-Nomological Model of Probabilistic Explanation." *Philosophy of Science* 45:206–226.

1981 "Probability, Explanation, and Information." *Synthese* 48:233–256.

Reichenbach, Hans

1944 *Philosophic Foundations of Quantum Mechanics*. Berkeley and Los Angeles: University of California Press.

1949 *The Theory of Probability*. Berkeley and Los Angeles: University of California Press.

1954 *Nomological Statements and Admissible Operations*. Amsterdam: North-Holland. Reprinted, with a new Foreword by Wesley C. Salmon, as (Reichenbach, 1976).

1956 *The Direction of Time*. Berkeley and Los Angeles: University of California Press.

1976 *Laws, Modalities, and Counterfactuals*. Berkeley and Los Angeles: University of California Press. Reprint of (Reichenbach, 1954).

Rescher, Nicholas

1970 *Scientific Explanation*. New York: Free Press.

Rogers, Ben
    1981   "Probabilistic Causality, Explanation, and Detection." *Synthese* 48:201–223.

Rosen, Deborah A.
    1975   "An Argument for the Logical Notion of a Memory Trace." *Philosophy of Science* 42:1–10.

Rothman, Milton A.
    1960   "Things That Go Faster Than Light." *Scientific American* 203, no. 1 (July):142–152.

Russell, Bertrand
    1927   *The Analysis of Matter*. London: George Allen and Unwin.
    1929   *Mysticism and Logic*. New York: W. W. Norton.
    1948   *Human Knowledge, Its Scope and Limits*. New York: Simon and Schuster.

Salmon, Wesley C.
    1959   "Psychoanalytic Theory and Evidence." In Sidney Hook, ed., *Psychoanalysis, Scientific Method, and Philosophy*, 252–267. New York: New York University Press.
    1963   "On Vindicating Induction." *Philosophy of Science* 30:252–261. Also published in Henry E. Kyburg, Jr., and Ernest Nagel, eds., *Induction: Some Current Issues*. Middletown, Conn.: Wesleyan University Press.
    1965   "The Status of Prior Probabilities in Statistical Explanation." *Philosophy of Science* 32:137–146.
    1965a   "Consistency, Transitivity, and Inductive Support." *Ratio* 7:164–169.
    1966   "The Foundations of Scientific Inference." In Robert G. Colodny, ed., *Mind and Cosmos*, 135–275. Pittsburgh: University of Pittsburgh Press. Reprinted in (Salmon, 1967).
    1967   *The Foundations of Scientific Inference*. Pittsburgh: University of Pittsburgh Press.
    1968   "Who needs inductive acceptance rules?" In Imre Lakatos, ed., *The Problem of Inductive Logic*, 139–144. Amsterdam: North-Holland.
    1969   "Partial Entailment as a Basis for Inductive Logic." In Nicholas Rescher, ed., *Essays in Honor of Carl G. Hempel*, 47–82. Dordrecht: D. Reidel.
    1970   "Statistical Explanation." In Robert G. Colodny, ed., *The Nature and Function of Scientific Theories*, 173–231. Pittsburgh: University of Pittsburgh Press. Reprinted in (Salmon et al., 1971).
    1970a   *Zeno's Paradoxes*. Indianapolis, Ind.: Bobbs-Merrill.

1971 "Explanation and Relevance: Comments on James G. Greeno's 'Theoretical Entities in Statistical Explanation.' " In Roger C. Buck and Robert S. Cohen, eds., *PSA 1970*, 27–37. Dordrecht: D. Reidel.

1974 "Comments on 'Hempel's Ambiguity' by J. Alberto Coffa." *Synthese* 28:165–174.

1975 *Space, Time, and Motion: A Philosophical Introduction*. Encino, Calif: Dickenson. 2nd ed., Minneapolis: University of Minnesota Press, 1980.

1975a "Theoretical Explanation." In Stephan Körner, ed., *Explanation*, 118–145. Oxford: Basil Blackwell.

1975b "Reply to Comments." In Stephan Körner, ed., *Explanation*, 160–184. Oxford: Basil Blackwell.

1975c "Confirmation and Relevance." In Grover Maxwell and Robert M. Anderson, Jr., *Minnesota Studies in the Philosophy of Science*, 6:3–36. Minneapolis: University of Minnesota Press.

1976 "Foreword." In Hans Reichenbach, *Laws, Modalities, and Counterfactuals*, vii–xlii. Berkeley/Los Angeles/London: University of California Press.

1977 "Objectively Homogeneous Reference Classes." *Synthese* 36:399–414.

1977a "A Third Dogma of Empiricism." In Robert Butts and Jaakko Hintikka, eds. *Basic Problems in Methodology and Linguistics*, 149–166. Dordrecht: D. Reidel.

1977b "Hempel's Conception of Inductive Inference in Inductive-Statistical Explanation." *Philosophy of Science* 44:180–185.

1977c "An 'At-At' Theory of Causal Influence." *Philosophy of Science* 44:215–224.

1977d "Indeterminism and Epistemic Relativization." *Philosophy of Science* 44:199–202.

1978 "Why ask, 'Why?'?" *Proceedings and Addresses of the American Philosophical Association* 51:683–705. Reprinted in (Salmon, 1979a).

1979 "Propensities: A Discussion-Review." *Erkenntnis* 14:183–216.

1979a *Hans Reichenbach: Logical Empiricist*. Dordrecht: D. Reidel.

1980 "Probabilistic Causality." *Pacific Philosophical Quarterly* 61:50–74.

1982 "Causality: Production and Propagation." In Peter Asquith and Ronald N. Giere, eds., *PSA 1980*, 49–69. East Lansing, Mich.: Philosophy of Science Association.

1982a "Comets, Pollen, and Dreams: Some Reflections on Scientific Explanation." In Robert McLaughlin, ed., *What? Where? When? Why?* 155–178. Dordrecht: D. Reidel.

1982b "Further Reflections." In Robert McLaughlin, ed., *What? Where? When? Why?* 231–280. Dordrecht: D. Reidel.

1982c "Causality in Archaeological Explanation." In Colin Renfrew et al., eds., *Theory and Explanation in Archaeology*, 45–55. New York: Academic Press.

Salmon, Wesley C., with contributions by Jeffrey, Richard C., and Greeno, James G.

1971 *Statistical Explanation and Statistical Relevance*. Pittsburgh: University of Pittsburgh Press.

Sayre, Kenneth M.

1977 "Statistical Models of Causal Relations." *Philosophy of Science* 44:203–214.

Scadron, Michael

1979 *Advanced Quantum Theory*. New York: Springer-Verlag.

Schnorr, C. P.

1971 "A Unified Approach to the Definition of Random Sequences." *Mathematical Systems Theory* 5:246–258.

Schwinger, Julian, ed.,

1958 *Selected Papers on Quantum Electrodynamics*. New York: Dover Publications.

Scriven, Michael

1959 "Explanation and Prediction in Evolutionary Theory." *Science* 130:477–482.

1975 "Causation as Explanation." *Nous* 9:3–16.

Singer, Barry, and Benassi, Victor A.

1981 "Occult Beliefs." *American Scientist* 69, no. 1:49–55.

Smith, A. H.

1958 *The Mushroom Hunter's Guide*. Ann Arbor: University of Michigan Press.

Stegmüller, Wolfgang

1973 *Problems und Resultate der Wissenschaftstheorie und Analytischen Philosophie*, Band 4, Studienausgabe Teil E. Berlin/New York: Springer-Verlag.

Stein, Howard

1982 "On the Present State of the Philosophy of Quantum Mechanics." Address to the 1982 meeting of the Philosophy of Science Association.

Suppes, Patrick

1970 *A Probabilistic Theory of Causality*. Amsterdam: North-Holland.

Suppes, Patrick, and Zanotti, Mario.

1981 "When are Probabilistic Explanations Possible?" *Synthese* 48:191–199.

Taylor, Edwin F., and Wheeler, John A.
 1966   *Spacetime Physics*. San Francisco: W. H. Freeman.
Teller, Paul
 1974   "On Why-Questions." *Nous* 8:371–380.
Thomas, J/P
 1979   "Homogeneity Conditions on the Statistical Relevance Model
        of Explanation." *Philosophical Studies* 36:101–106.
Tuomela, Raimo
 1977   *Human Action and Its Explanation*. Dordrecht: D. Reidel.
 1981   "Inductive Explanation." *Synthese* 48:257–294.
Turing, A. M.
 1937   "Computability and λ-definability." *Journal of Symbolic Logic*
        2:153–163.
Vaihinger, Hans
 1924   *The Philosophy of 'As If.'* London: Kegan Paul.
van Fraassen, Bas C.
 1977   "The Pragmatics of Explanation." *American Philosophical
        Quarterly* 14:143–150.
 1980   *The Scientific Image*. Oxford: Clarendon Press.
 1982   "Rational Belief and the Common Cause Principle." In Robert
        McLaughlin, ed., *What? Where? When? Why?* 193–209. Dor-
        drecht: D. Reidel.
 1982a "The Charybdis of Realism: Epistemological Implications of
        Bell's Inequality." *Synthese* 52:25–38.
Venn, John
 1866   *The Logic of Chance*. London: Macmillan.
von Bretzel, Philip
 1977   "Concerning a Probabilistic Theory of Causation Adequate for
        the Causal Theory of Time." *Synthese* 35:173–190. Reprinted
        in (Salmon, 1979a).
von Mises, Richard
 1957   *Probability, Statistics and Truth*, 2nd rev. English ed. New York:
        Macmillan.
 1964   *Mathematical Theory of Probability and Statistics*. New York:
        Academic Press.
von Wright, Georg Henrik
 1971   *Explanation and Understanding*. Ithaca, N.Y.: Cornell Univer-
        sity Press.
Weinberg, Steven
 1972   *Gravitation and Cosmology*. New York: John Wiley and Sons.
 1979   "Einstein and Spacetime: Then and Now." *Bulletin of the Amer-
        ican Academy of Arts and Sciences* 33:35–47.

Wessels, Linda
   1982   "The Origins of Born's Statistical Interpretation." In P. Asquith
          and R. N. Giere, eds., *PSA 1980*, vol. 2:187–200. East Lansing,
          Mich.: Philosophy of Science Association.
Westfall, Richard S.
   1971   *The Construction of Modern Science: Mechanisms and Mechan-
          ics*. New York: John Wiley and Sons.
White, Robert L.
   1966   *Basic Quantum Mechanics*. New York: McGraw-Hill.
Wichmann, Eyvind H.
   1967   *Quantum Physics* (Berkeley Physics Course, vol. 4). New York:
          McGraw-Hill.
Winnie, John
   1977   "The Causal Theory of Space-time." In John Earman, Clark
          Glymour, and John Stachel, eds., *Minnesota Studies in the Phi-
          losophy of Science*, 8:134–205. Minneapolis: University of Min-
          nesota Press.
Wright, Larry
   1976   *Teleological Explanation*. Berkeley and Los Angeles: University
          of California Press.

Index

acceptance, rule of, 89-90, 191
Achinstein, Peter, 22n, 240, 273
action-at-a-distance, 209, 242, 244-245, 250, 255, 258; in quantum mechanics, 210
action-by-contact, 210, 242
adequacy: condition of, 113; empirical, 134n, 229, 238
admissible selective class, *see* selective class
agreement (values of N), 219-221, 237
air, humidity-density relation, 268-269, 273
aleatory explanation, 34n, 265-267
alpha radiation, 217, 221, 224-226, 236
alpha scattering, 119, 203, 262
ambiguity, *see* inductive-statistical model, ambiguity of
Anderson, Alan Ross, 93
anomaly, causal, 229, 243-244, 255
answer, to why-question, 102-107, 110; core of, 104-105; direct, 103-105, 108
anthropomorphism, 8, 13
Aquinas, St. Thomas, 26, 217
argument: analogical, 233-237; causal, 233; common cause, 223
Aristotle, 4, 18, 18n, 20, 21, 135
Arnauld, Antoine, 4, 21
'as if,' philosophy of, 238
associated sequence, 61, 64, 69, 71, 75-79; selection by, 61-64, 70, 76, 79-81
Astrodome (fictitious), 141-147 *passim*
astrology, 26
asymmetry, temporal, 72, 94-96, 163-164, 166, 175-176
asymmetry thesis (Reichenbach), 163, 166
at-at theory, 147-155
atom: as hypothetical object, 216, 219; decay from excited state, 200; marking of, 202; positronium, 246-247, 250-251; quantized model of, 228-229; spontaneous swerving of, 25; tritium, 277

atomic/molecular hypothesis, *see* atomism
atomism, 5-6, 214-215, 219, 241; ancient, 214; argument for, 223; Dalton's, 214; experimental proof of, 216, 228; opponents of, 251; rejection of, 221-222
Avogadro's law, 228, 269
Avogadro's number, N, 180, 214-224, 226, 228, 236-237, 251

background conditions (for conjunctive fork), 162-163, 169, 175, 179, 223, 227
Balmer's formula, 6, 252-253
barometer/storm, 13, 43-44, 127, 211
basic entities: events as, 138; processes as, 139-140
Bayesianism, objective, 234
Bayes's theorem, 205, 234-235
becoming, cinematographic view of, 151
behaviorism, 7
Bell's inequality, 250, 254
Belnap, Nuel D., Jr., 93, 102
Benassi, Victor A., 12n
Bergson, Henri, 151, 153n
Berkeley, George, 231
Bernoulli principle, 187
betweenness, causal, 192, 194, 200
billiard balls, collision of, 136-137. *See also* pool balls
Bizet, Georges, 149
blackbody radiation, 216, 218, 221, 228
Bohm, David, 246, 257
Bohr, Niels, 228-229, 243-245
Bohr orbits, 253
Bolton, Herbert E., 11
Boltzmann's constant, 218
bomb, atomic, 270-271
bone, worked, 267-270, 273
bookkeeping, sloppy, 257-258
Born, Max, 26, 118, 255
Braithwaite, R. B., 18, 21
branch structure, hypothesis of, 180
Bridgman, Percy W., 14n

Library of Congress Cataloging in Publication Data

Salmon, Wesley C.
  Scientific explanation and the causal structure of the world.

  Bibliography: p.
  Includes index.
  1. Science—Philosophy. 2. Science—Methodology.
I. Title.
Q175.S23415 1984   501   84-42562
ISBN 0-691-07293-0 (alk. paper)
ISBN 0-691-10170-1 (pbk.: alk. paper)

Wesley C. Salmon is University Professor in the Department of Philosophy at the
University of Pittsburgh. Among his other books are Logic (Prentice-Hall, 1963),
The Foundations of Scientific Inference (Pittsburgh, 1967), Statistical Explanation
and Statistical Relevance (Pittsburgh, 1971), and Space, Time, and Motion (Dickenson, 1975).